# Memorizing Pharmacology

## A Relaxed Approach

Tony Guerra, Pharm.D.

Memorizing Pharmacology: A Relaxed Approach
Tony Guerra, Pharm.D.
First Edition
ISBN: 978-1-329-89844-8

*To Mindy,*
*Brielle, Rianne, and Teagan*

# AUTHOR'S NOTE

## THE STORY OF PHARMACOLOGY CLASS

As an instructor, I didn't appreciate how tough it was to be a working parent and pharmacology student until I had my triplet daughters. When the girls originally came home, they had trouble coordinating the suck, swallow, breathe that goes with feeding from a bottle, well, three bottles. We fed them for 90 minutes and they slept for 90 minutes – round the clock. They got older, but the crushing emotional and physical exhaustion has continued. Most of my students have jobs and families. I wanted a way for them to study pharmacology while attending to these types of responsibilities.

Tonight, as I lay quietly in bed next to my daughter who will not sleep otherwise, **I recited in my head, from memory, the 200 drugs, generic name, brand name and drug classification, in this book, in order.** Previously during these times, thoughts of "What do I have to do for tomorrow?" ran through my head. Instead, I could relax, stay unhurriedly by my daughter, and remain a committed parent. With this book's techniques, I could study pharmacology in the dark – eyes closed. I knew I was on to something. This could help my overworked students.

After she fell asleep, I got up and starting writing this introduction, but before I could congratulate myself, my other daughter came downstairs insisting that she will never sleep in her bed again. So, here I am, on my couch, half-watching *Better Call Saul*, comforting another daughter between coughing fits, and typing out this introduction, knowing I might see the sunrise before I get to sleep.

Earlier this fall a student emailed me, offering to pay me to read and record his "Top 200 Drugs" list. I was too busy mid-semester, but I did want to create a road map for him and other students that shared his need. Each fall and spring, many health professional students try to learn the "Top 200 Drugs," but get little help in memorization techniques.

This book *will* help, regardless of which Top 200 Drugs your professor assigns. It will also help you answer the question: where do I start studying pharmacology for my board exams? Whether it's the NCLEX-RN (National Council Licensure Examination - Registered Nurse) or NABPLEX (National Association of Boards of Pharmacy Licensure Examination) or another exam, this book will provide a structure for your knowledge to wrap around, much as a garden lattice might hold up vines.

What does Top 200 Drugs mean? At one time, there was a website that listed The Top 200 Drugs in the United States by 1) Number of prescriptions written and 2) Ranked by money spent on each drug. This created two different lists, but their value was clear. **Well-prepared students remember the most frequently prescribed drugs.** The 80/20 rule, or Pareto's principle, predicts that 20% of the medications will represent 80% of those prescribed.

Brutal rote memorization, however, is a poor strategy for managing information overload. Memorizing a brand name, generic name, medication class, therapeutic use, and one adverse effect for 200 drugs represents one-thousand pieces of information. Each new item adds 200 disparate memorization points. Students often ask why they should memorize drugs if they can Google them or look them up in a *Davis Drug Guide for Nurses*. The answer: a properly sorted and memorized list of 200 drugs provides a framework on which to build your preparation for pharmacology class, the board exams, and clinical practice. But, how did *I* memorize so many drugs? Follow this thought experiment.

Imagine, instead of playing the part of student, you are the instructor. The class has 200 students and there are exactly 200 seats in the classroom. How do you remember all of their names? How do you know who is absent when you see empty chairs? You could ask the students to sit alphabetically by first name or by last name, but this is college, not grade school. A more organic approach would take time to understand why groups formed as they did.

Students sit in the same seats weekly. As you talk and get to know them, you find out what brings them together. They may have the same undergraduate major, hometown, dorm, previous class, and so forth. There are the front row groups that always asks questions or back row group that asks no questions. Groups come from the same hospital floor or work on a semester project. If someone's absent, you know what group he or she are missing from.

As you learn the drugs in this book, you'll see similar groupings. For example, you'll see thirteen gastrointestinal drugs, five SSRI antidepressants, four benzodiazepines, and three angiotensin-converting enzyme inhibitors. When you bring medicines back from memory, you'll remember if one is missing.

This book provides that specific framework to memorize the most important drugs in a logical order. A good analogy is that learning to drive a car requires essentially the same foundational instruction. However, once you learn, you can drive anywhere (or anything) you want.

After you learn *this* list, you can insert and delete drugs. If you just want to know how to sort that brick of note cards you've got working as a doorstop right now, you can jump to the sorted list of 350 drugs in Chapter 9. I did not put 350 drugs in it because I expect you to memorize that many; rather, I put that many in there because the Top 200 lists from college to college can be so different.

**We have a problem, however, because pharmacology instructors and students speak different languages.** This causes instructors to get student comments like "He can't teach," "I didn't learn anything in his class," and "C's get degrees, I guess." Professors who cannot connect to students frustrate those who want to learn. I am especially empathetic to parents and students with full-time jobs and international students trying to pick up English *and* the language of pharmacology. I am a parent of three daughters and English is not my first language. Let me show you this disconnect between a student's and a teacher's memorization metaphors.

*Pharmacology* breaks down to "pharmaco" or "drug" and "logy" which is "study of" with a connecting "o" to make "study of drugs." There is another concept called *pharmacokinetics*. "Pharmaco" means "drug" and "kinetics" means movement, like a kinesiology major is someone who studies – "logy" – body movement – "kinesio."

When, as a pharmacology instructor, I look at the word *pharmacokinetics*, I think of four major principles: *absorption, distribution, metabolism*, and *excretion*. I remember the A-D-M-E mnemonic, pronounced: "add-me." I remember each principle as matching a certain organ or tissue.

- Absorption happens in the *small intestine's* villi.
- Distribution occurs via the *blood.*
- Metabolism happens in the *liver.*
- Excretion occurs through the *kidneys.*

With my anatomy and physiology background, I memorized these facts through a metaphor of connected tunnels. A drug travels from the mouth to the small intestine to be *absorbed* and then through blood vessels to be *distributed* to the liver where it's *metabolized*. Eventually it goes through the kidneys to be *excreted*, further passing through the ureters, into the bladder, and the urethra. That picture is clear and complete in my brain.

**I used to assume that my students, some of whom took anatomy and physiology, learned it the same way.** That wasn't true. When asked to create a writing prompt of how they memorized the same idea, my students used a different set of metaphors.

One equated *absorption* with soaking up material in a classroom, *distributing* knowledge to short- and long-term memory, *metabolizing* the information down into manageable topics, and *excretion* to eliminating extraneous pieces of information.

Another student condensed a semester's worth of developmental psychology into a sentence. She associated the increased surface area of her pregnant belly with the intestine's *absorption; distribution* with the bloodline that she imparts to her child; changes in her life to accommodate the child, including possibly future drinking after pregnancy, to remember the *metabolism* in the liver; and *excretion* as the expulsion of the child from her home after high school.

After reading that story, can you tell me what *pharmacology* and *pharmacokinetics* mean? Whose story did you use, the instructor's or the student's? I don't think one is better than the other is, but it can be helpful to have a more complete understanding through two points of view.

## THE CURSE OF KNOWLEDGE

This divide between expert and novice has a name – "The Curse of Knowledge." When you have been an expert at something, but you just can't explain it to someone, it might be because you understand the whole process and they are just learning step-by-step. The same is true in pharm.

Whatever pharmacology text you are using has an author trying to tell the same story from a different expert point of view to novice audiences. I've found pharmacology books specifically titled for allied health, athletic training, audiologists, dental hygienists, EMS providers, health professions, massage therapists, medical assistants, medical office workers, nurses, paramedics, pharmacists, pharmacy technicians, physical therapists, physicians, rehabilitation professionals, respiratory care therapists, surgical technologists, veterinary technicians and veterinarians. I stopped counting at twenty. Even as someone writes a textbook, the base of knowledge expands and changes. The better approach starts with a general primer like this, a very small amount of information, 200 drugs in this case, and expands from it. Then a student can take what he or she specifically needs from discipline-specific pharmacology textbooks.

I have read many of those discipline-specific books and taught in many disciplines, but you get the point – **pharmacology is an ever-expanding subject too big to distill; it's best learned by acquiring a base and building up from there.** I would recommend that you find out how your fellow students learn, but in a big lecture hall, you often don't talk to classmates. This book picks up not only the shortcuts those students use to help them get a better grade in pharm, but also their methods.

## TAKING PHARMACOLOGY TWICE, SORT OF.

Most of the above-mentioned pharmacology textbooks have the same basic format. The early chapters cover the interactions of the drugs with the body, *pharmacodynamics*, and the body on the drug, *pharmacokinetics*, along with introducing some basic vocabulary. The author divides chapters by pathophysiologic condition and medications follow, e.g., gastrointestinal medicines, then cardiac medicines, and so forth.

However, many sections rely on understanding material from a future section. The instructor has a full overview of all the knowledge, but the student doesn't. Soon, the instructor is talking over the student's head. Here is an example: What medications treat an ulcer and why?

A student can memorize the three medications, **omeprazole (Prilosec), amoxicillin (Amoxil),** and **clarithromycin (Biaxin)** that comprise an ulcer treatment regimen. However, to move up Bloom's taxonomy from knowledge (memorizing "what") to comprehension (understanding "why") requires significant additional information.

- **Omeprazole (Prilosec)**, a proton pump inhibitor, reduces stomach acid and gains an advantage over **calcium carbonate (Tums**) and **ranitidine (Zantac)** because of its extended half-life, a measure of the time it takes for a drug in the body to reduce by half.

- **Amoxicillin (Amoxil)** kills the causative agent in most ulcers, *Helicobacter pylori,* a helicopter- shaped bacterium that's sensitive to penicillin antibiotics.

- **Clarithromycin (Biaxin)** reduces the incidence of resistance that can happen with a single broad-spectrum antibiotic like **amoxicillin (Amoxil).**

Instructors teach **omeprazole (Prilosec)** in the earlier gastrointestinal section. Weeks later, students learn how **amoxicillin (Amoxil)** and **clarithromycin (Biaxin)** work in the antimicrobial section of the course. This second major point is critical – **To learn pharmacology requires that a student has already taken pharmacology.** This seems ridiculous, but let's use an apt metaphor, a trip to another country.

## CREATING A PRIMER

Imagine you will start college in the fall and have decided to take a four-credit Spanish class. You could hope the teacher slows down to whatever speed you need, or you could study ahead of time. Let's pretend you decide to travel to the Peruvian mountains. (That happens to be where my dad is from.) Although some people speak English, you ask them to speak to you in Spanish so they can help you learn Spanish for your class. Your smartphone doesn't get any signal in the Andes Mountains, so you use pen and paper to write your notes in a small journal, picking up words along the way. At first, you point to objects but, gradually, you can start making sentences. You make notes of words that are especially tricky. For example, the Spanish word embarazada *looks like* the English word embarrassed, but it *means* pregnant. If you make this mistake, your brain will remember the story of that mistake, and you won't make it again. You compare notes with other students and share stories each night. In the end, you have a primer, an introductory book in Spanish that will put you far ahead of your class as you enter college.

The word "primer" has different meanings in various contexts. In auto painting and cosmetics, a primer allows for better adhesion for a secondary product's application. A primer is also a basic language-reading textbook. You can combine these definitions to create a useful analogy. **A pharmacology primer can provide a home base for students to expand a basic framework of vocabulary before they use PowerPoints (sentences) and textbooks (full paragraphs).**

This book is your journal. In the same way you take medical terminology before you take anatomy and physiology, you want to master some of the terms before you start pharm class. The following seven chapters include conversations with students about creating meaning from words in the foreign language of pharmacology. I will talk through how and why I have grouped them as I did. Once you have completed the book, the Comprehensive Drug List should look like conversational and readily pronounceable English.

## Medication reconciliation (Med Rec)

While this book's focus is to prepare students, I believe it can help patients and caregivers provide accurate medication histories and records - as students of their own conditions. Medical reconciliation or "med rec" is the process of assembling a correct list of patient's medications under what might be some stressful circumstances. When an ambulance comes, you might not remember to take everything with you. Memorizing medications in a logical order ensures the list will always be with you and you can immediately tell EMS responders what the patient is on.

## Final note

If you invent a great mnemonic and want to share it for a future edition, feel free to email me:

**aaguerra@dmacc.edu**

I'm always looking for better ways to teach pharm. If you see a student struggling with the language of pharmacology who doesn't have this book or a patient who wants to better understand his or her many prescriptions, please do take the time to recommend it to them.

There was a pharmacology class that I got a standing ovation, just like in the 1972 movie *The Paper Chase*. I hope what you get out of this book is worth your standing up.

Tony Guerra
December 28th, 2015
2:41 AM - Still dark outside.

# Table of Contents

Contents

# INTRODUCTION

## GENERIC NAMES VS. BRAND NAMES

From experience, instructors immediately know which names represent *generic* and *brand* drugs. In an academic text, generic names go first in lower case letters followed by a capitalized brand name in parenthesis, e.g., **amphotericin B (Fungizone)**. In conversation or other writing, students must have this distinction memorized.

Some licensing exams only use generic names. This is understandable; only one generic name exists for each medication. There is one "**acetaminophen**," but the brand **Tylenol** is associated with many products. This *does not* mean a student should *not learn* brand names. A dangerous confusion between generic and brand names is that of **Pepto-Bismol** or **Pepto** for short. Let me tell you a story.

On a web page I saw a parent post that she gave her 8-year-old child a teaspoonful of **Pepto**, was that okay? **Pepto-Bismol** is the brand name for the liquid product **bismuth subsalicylate**. That salicylate is similar to **aspirin**, which is **acetylsalicylic acid** and can cause a terrible condition called **Reye's syndrome** in children. What she meant to give her child was **Children's Pepto**, which contains **calcium carbonate**, the active ingredient in **Tums**. As a parent, I understand what happens in the middle of the night. Your child is suffering and the medication instructions are in tiny 4-point font. It's tough to make the right call.

The point is that most brand names will have two to three syllables (**Pepto**). Most generic names have four or more: **calcium carbonate** and **bismuth subsalicylate**. Why is this important?

You want to start developing simple *heuristics,* or rules of thumb, to make it easier to learn the medicines. For example, generic names have stems like the "-cillin" in penicillin class antibiotics; brand names do not. Getting meaning from a brand or generic requires a different rule of thumb. If you don't know if the drug name you are looking at is generic or brand, you are at a terrible disadvantage.

## THREE TYPES OF DRUG NAMES

**1) The chemical name:** First, there is the most complex name, the International Union of Pure and Applied Chemistry (IUPAC) standard name, which makes perfect sense to a chemist who might want to draw the molecule. For example, **Ibuprofen's** chemical name is:

(RS)-2-(4-(2-methylpropyl)phenyl) propanoic acid

Chemists may simplify the chemical name. In this example, the chemical name becomes the common name, Iso-butyl-propanoic-phenolic acid (I bu pro phen), which can be shortened even further to ibuprofen, the generic name.

**2) The generic name:** With only four syllables, the transformation to **ibuprofen** is an improvement over the chemical name. Patients, however, prefer two to three syllable names. Just as kids cut down computer to "puter" or banana to "nana," patients prefer short names for ease of pronunciation.

If you don't know where to put an emphasis on a generic four syllable drug name, it's usually the penultimate, or second to last, so it would be pronounced **i bu PRO fen**.

You want to make sure to put the right *emphasis* on the right *syllable* or you'll lose credibility. If your patient pushes the call button when you walk *into* the room, you might be mispronouncing some meds.

**3) The brand or trade name:** Two of the brand names for **ibuprofen** are **Advil** and **Motrin**. Both have two easy-to-pronounce syllables, but don't resemble **ibuprofen** as the brand name because they include *plosives* to make powerful memorable stops in the word. A strong word sounds like strong medicine.

Say each of the following letters and see if you can feel it in your tongue or nose.

Tongue blade occlusion: *t or d*

Tongue body occlusion: *k or g*

Lip occlusion: *b or p*

Nasal stops: *m or n*

**Motrin** has an "m," a nasal stop; a "t," a tongue blade occlusion; and an "n," another nasal stop. This forces the person saying the word to stop their breath three times, slowing the pronunciation and keeping it on the tongue and / or nose longer, making it sound very strong.

## BRAND NAME RULE OF THUMB: CAN INDICATE FUNCTION

The **brand name's** two to three syllables will sometimes hint at the *function* of a drug, i.e., **Lopressor** lowers blood pressure

The **brand name** is similar to a two- to three-syllable nickname that *hints* at the drug's function, but by law, may not make a claim. Brand names are very much like nicknames such as Betsy or Jack. A non-native English speaker would have no idea that Betsy comes from Elizabeth or Jack comes from Jonathan. Betsy takes the "b-e-t" from Elizabeth and Jack takes the "J" from Jonathan.

You might see brand names do the same, such as the use of "val" in the brand name **Valtrex** (an antiviral), which comes from a part of the generic **valacyclovir** or the "p," "a," and "x" in the brand name **Paxil** (an antidepressant), which come from the generic name paroxetine.

## GENERIC NAME RULE OF THUMB: CAN INDICATE CLASS

Lopressor's four syllable generic name, **metoprolol (Toprol, Toprol XL)**, has an -olol suffix, which is a **stem** from a credible source.

I know of two stem lists. One is from the **United States Adopted Names Council** on the American Medical Association's website and another is the **World Health Organization's** (WHO) Stem Book. They created stems so when people use generic names, they will know that the drugs with similar stems are probably similar in their actions. This is where the problems start.

Students, recognizing similarities in the endings or beginnings of drug names, start to make up their own rules. In this book, you will learn which stems are credible and verifiable and which are not. Knowing both will make you much better at spotting them. Most YouTube videos and Quizlet notecards have errors that let you know the author's research is faulty or nonexistent. I am not picking on students; many licensed health professionals give advice and charge money online to view their videos, and some of that advice is dead wrong. However, when used correctly, these prefixes, suffixes, and infixes can be invaluable.

Some of the references, online videos, and online pre-made note cards provide lists deriving stems incorrectly. For example, the nine endings -azole, -en, -ide, -in, -ine, -one, -pam, -sone, and -zine are *not* stems; these are only groups of letters that happen to be at the end of many medication names. Using a group of letters instead of an established stem might lead to a drug classification error.

## PREFIXES, SUFFIXES, AND INFIXES

In this book, I will use the terms prefix, suffix, and infix. Generic drug names are invented words and do not always conform to the rules of English. The United States Adopted Names Council properly calls each prefix and suffix that has meaning (e.g., cef- represents cephalosporins or –cillin represents penicillins) a stem.

In the case of penicillin, -cillin is the stem and peni- is a prefix that differentiates peni*cillin* from other penicillin antibiotics such as amoxi*cillin* or ampi*cillin*, etc.

An infix is inside the word to make the classification more specific. The proper stem for a quinolone antibiotic is –oxacin, but cipro*floxacin* has the infix *–fl-* to classify it further as a *fluoro*quinolone, one that contains a fluorine atom.

Stems work as a heuristic, or something that allows us to accelerate our recognition of a class of medications. If there is a list of ten medications that all end in –olol, we can more easily know these medications are beta-blockers.

The stem –adol indicates a drug like **tramadol (Ultram)**, is an analgesic (a medication for pain), comprised of an opiate (which means from the opium poppy, but is more generally a term for many narcotics). This has properties as both an agonist (a chemical that stimulates a receptor) and an antagonist (a chemical that blocks or antagonizes a receptor), which is unusual. Usually a chemical is an agonist or antagonist, not both.

The stem -afil in **sildenafil (Viagra)** and **tadalafil (Cialis)** represents the phosphodiesterase type-5 (PDE5) inhibitors and indicates that these medications block (inhibit) an enzyme (phosphodiesterase). The inhibition of this enzyme results in stopping the breakdown of a chemical in the corpus cavernosum. This effect helps patients who have erectile dysfunction.

The stem –amivir in **oseltamivir (Tamiflu)** and **zanamivir (Relenza)** is a subclass of the stem –vir that represents the neuraminidase (enzyme) antagonist (inhibitor) group. Neuraminidase is an enzyme critical for influenza virus replication. If the medication in given in a certain time frame, usually within the first 48 hours after symptom onset, it blocks the enzymes necessary for influenza viruses to successfully replicate.

The stem "-azepam" represents antianxiety agents in the benzodiazepine class that are similar to **di<u>azepam</u> (Valium)**. Besides anxiety, patients use these medications as a sedative-hypnotic.

That's a lot of information to get from a few letters.

## PRONUNCIATION AND ORGANIC CHEMISTRY WORD PARTS

Pronouncing generic names is often like pronouncing foreign language last names like mine. My last name "Guerra" has a part that's unpronounceable in regular English because there is no double or rolled "r" sound in English. Drug names might use the same letters you know in the Roman alphabet. However, we pronounce them differently because the sounds that form them come from organic chemists and biochemists.

Note: Below I have italicized the part of the generic name to which the organic molecule corresponds. These are *not* classification stems like -cillin or -azepam, but references to certain chemicals or groups of them.

*These words indicate the number of carbon atoms in an attached molecule made up of only carbon and hydrogen atoms:*

| | |
|---|---|
| **Methyl** - *Methyl*phenidate | *METH-ill* |
| **Ethyl** - Fentan*yl* | *ETH-ill* |
| **Propyl** - Meto*pro*lol | *PROP-ill* |
| **Butyl** - Al*but*erol | *BYOOT-ill* |

*Levo and dextro mean left and right respectively:*

| | |
|---|---|
| **Levo** - *Levo*thyroxine | *LEE-vo* |
| **Dextro** - *Dex*methylphenidate | *DEX-trow* |

*These words mean there is a specific element in each molecule:*

| | | |
|---|---|---|
| **Thio** (sulfur) - Hydrochloro*thi*azide | | *THIGH-oh* |
| **Chloro** (chlorine) - Hydro*chloro*thiazide | | *KLOR-oh* |
| **Hydro** (hydrogen) - *Hydro*codone | | *HIGH-droe* |
| **Fluoro** (fluorine) - Cipro*fl*oxacin | | *FLOR-oh* |

*These words are branches that attach to the central molecule:*

| | |
|---|---|
| **Acetyl** - Levetir*acet*am | *Uh-SEAT-ill* |
| **Alcohol** - Tramad*ol* | *AL-kuh-haul* |
| **Amide** - Loper*amide* | *UH-myde* |
| **Amine** - Diphenhydr*amine* | *UH-mean* |
| **Disulfide** - *Disulf*iram | *DIE-sulf-eyed* |
| **Furan** - *Fur*osemide | *FYOOR-an* |
| **Guanidine** - Cimet*idine* | *GWAN-eh-dean* |
| **Hydroxide** - Magnesium *Hydroxide* | *HI-drox-eyed* |
| **Imidazole** - Omepr*azole* | *im-id-AZ-ole* |
| **Ketone** - Spironolact*one* | *KEY-tone* |
| **Phenol** - Acetamino*phen* | *FEN-ole* |
| **Sulfa** - *Sulfa*methoxazole | *SULL-fuh* |

Some chemists name drugs:

- By what they do for the patient, also called the therapeutic class: Anti-depressant
- By their chemical structure: Tricyclic antidepressants (TCAs) (three rings in the compound)
- By the receptor they affect: Beta-blockers
- By the neurotransmitter they affect: Selective serotonin reuptake inhibitor (SSRI)

You will become familiar with these classifications as we progress through the book. That's another big reason you want to be around other people when discussing the medications. You will pick up the pronunciation as you make mistakes or listen to others make mistakes. There is no shame in this; it's a part of learning. You cannot exactly pick up pronunciation from looking at brand and generic names. However, if you pay attention to drug names, you will find clues for building strong mnemonics.

## HOMOPHONES

In grade school, you may have learned the word "homophone." Homophones are words that sound the same, but that we spell differently. Some examples include the words "there" and "their," "two" and "too," and "hear" and "here." To remember which meaning is associated with the word, your teacher may have told you to look inside the word for a clue.

In the word "their" you find "heir," such as the person who will inherit something. Then you can associate that the word "their" has to do with the possessive form of a group of people versus "there" which means "in that place."

In the word "two," you can turn the "w" sideways to make a 3, spelling "t3o" to remember that two has to do with a number. Also, "too" has two o's, and you can remember that it has too many o's.

In the word "hear" you find the word "ear," which reminds you this word means to listen, versus "here" which means "in this place."

Those clues are mnemonic devices, something to help your memory. Mnemonics also work in the memorization of drugs.

Note: The word mnemonic comes from the name Mnemosyne, the goddess of memory in Greek mythology.

## Mnemonics in Pharmacology

The brand name **Prilosec**, a proton pump inhibitor for reducing stomach acid, contains "Pr" which can be short for "proton" ($H^+$), the ion associated with something acidic. **Prilosec** contains "l-o" which can be short for "low" as in the opposite of high. **Prilosec** contains "sec" which can be short for "secretion."

**Prilosec's** mechanism of action (MOA) is to inhibit proton pumps and reduce the acid in a person's stomach. By looking at the name of the drug, we can see that "proton" "low" "secretion" means a reduction in protons, helping us remember the meaning of the word.

However, if you only remembered the generic name, **omeprazole**, and forgot what the stem -prazole meant or what the drug was for, you would be in trouble. Brand names serve as a back-up plan.

When developing my mnemonics, I did not call anyone at any brand name drug companies. I just looked at each drug name and used my experience as a teacher of pathophys, pharmacology, and organic and biochemistry, and made up something that seems to make sense, but that, more importantly, will help students remember the drug's drug class and / or function. The FDA does not allow a drug company to name a drug after its intended use, but there are hints in many drug names that you can definitely see.

## 3 BY 5 NOTECARDS

Many students prefer to use notecards rather than just relying on a book. I think you can get everything you need from this book alone, but I know that 3 by 5 notecards you *make* are much better than any you can *buy*. A book from Harvard University Press titled *Make It Stick: The Science of Successful Learning*, by Peter Brown, goes into why generative (making things) learning is so important.

Notecards are portable. You can sort them and challenge others to sort them like UNO or playing cards. To help you along, I have created a specific order that makes sense out of these 200 cards so you know *how* to sort them. The "how to sort them" aspect is the next level of learning beyond memorizing the purpose of drugs or their generic and brand names.

Drugs next to each other are related and those within a group are in a larger family based on physiologic system. As you go through the book, you will see connections between the drug before and after the one you are studying.

You will also see what *types* of connections are available in addition to those you know already. We will take small steps, but I know you will be impressed when you can name every single drug's brand name, generic name, and class from memory.

Here is what a notecard might look like:

**3x5 card front:**
famotidine

**3x5 card back:**
**Class:** $H_2$ blockers have the stem "-tidine." Looks like "to dine," which can be associated with GERD.

The brand name **Pepcid** has the "pep" from peptic, which means digestion, and the "cid" from acid.

## COMPREHENSIVE DRUG LIST DISCUSSION

I didn't make this three-page list to intimidate you. It's just that if you spread out 200 3 x 5 notecards, 10 down and 20 across, it makes an area 4 feet by 8 feet. This list helps you see the whole forest – connections between drugs and stems unapparent in a stack of cards.

Start by memorizing the seven pathophysiologic classes in this book in order as G-M-RINCE, as Grand Mothers RINCE kids' hair (except it's the French r-i-n-c-e instead of the English r-i-n-s-e) to set up the broadest framework. These seven pathophysiologic classes will be the steel reinforcing bars that, when surrounded by concrete, will provide the foundation for your memorizing the gastrointestinal (G), musculoskeletal (M), respiratory (R), immune (I), neuro (N), cardio (C), and endocrine (E) systems' medications.

This G-M-RINCE order places drug classes from easiest to hardest to learn. To make it easier to memorize the whole thing, I had to create some simple rules – a computer programmer might call these algorithms. In each section, these are the two major rules:

1. Each of the seven sections has over-the-counter medications presented first and then prescription medications after. I present all OTC gastrointestinal products before all RX products, so the drugs consumers can physically interact with at the pharmacy come first.

2. I alphabetized drugs in the same class unless there is a pharmacologic reason to consider them out of alphabetical order.

For example, **diphenhydramine**, a 1st-generation antihistamine that starts with "d," would go before **cetirizine**, a 2nd-generation antihistamine that starts with "c." The generational move from 1st to 2nd overrides the alphabetical order. However, **cetirizine** and **loratadine** are both 2nd generation antihistamines so those *are* in alphabetical order. Therefore, the order becomes **diphenhydramine, cetirizine, loratadine** - one first-generation and two second-generation antihistamines.

This is how our brains work - ever consolidating and organizing until meaning emerges from a compact, somewhat fractured list. The ultimate goal is to consolidate all seven chapters in the comprehensive drug list. You shouldn't need 200 notecards to memorize 200 drugs; you should only need 7 - one for each physiologic group.

# COMPREHENSIVE DRUG LIST

(**Bolded Drugs** generally do not require a prescription)

## Chapter 1 – Gastrointestinal

| | | |
|---|---|---|
| **Calcium carbonate** | **Bismuth subsalicylate** | Infliximab |
| **Magnesium hydroxide** | **Loperamide** | |
| **Famotidine** | **Docusate sodium** | |
| **Ranitidine** | **Polyethylene glycol** | |
| **Esomeprazole** | Ondansetron | |
| **Omeprazole** | Promethazine | |

## Chapter 2 – Musculoskeletal

| | | |
|---|---|---|
| **Aspirin** | Morphine | Methotrexate |
| **Ibuprofen** | Fentanyl | Abatacept |
| **Naproxen** | Hydrocodone/APAP | Etanercept |
| **Acetaminophen** | Oxycodone/APAP | Alendronate |
| **ASA/APAP/Caffeine** | APAP/Codeine | Ibandronate |
| Meloxicam | Tramadol | Cyclobenzaprine |
| Celecoxib | Naloxone | Diazepam |
| | Eletriptan | Allopurinol |
| | Sumatriptan | Febuxostat |

## Chapter 3 – Respiratory

| | | |
|---|---|---|
| **Diphenhydramine** | **Triamcinolone** | Fluticasone |
| **Cetirizine** | **Guaifenesin/DM** | Albuterol |
| **Loratadine** | Guaifenesin/codeine | Albuterol/Ipratropium |
| **Loratadine-D** | Methylprednisolone | Tiotropium |
| **Pseudoephedrine** | Prednisone | Montelukast |
| **Phenylephrine** | Budesonide/Formoterol | Omalizumab |
| **Oxymetazoline** | Fluticasone/Salmeterol | Epinephrine |

# Chapter 4 - Immune

- **Neomycin / Polymyxin-B / Bacitracin**
- **Butenafine**
- **Influenza vaccine**
- **Docosanol**
- Amoxicillin
- Amoxicillin / Clavulanate
- Cephalexin
- Ceftriaxone
- Cefepime
- Vancomycin
- Doxycycline
- Minocycline
- Azithromycin
- Clarithromycin
- Erythromycin
- Clindamycin
- Linezolid
- Amikacin
- Gentamicin
- Sulfamethoxazole / Trimethoprim
- Ciprofloxacin
- Levofloxacin
- Metronidazole
- Rifampin
- Isoniazid (INH)
- Pyrazinamide (PZA)
- Ethambutol
- Amphotericin B
- Fluconazole
- Nystatin
- Oseltamivir
- Zanamivir
- Acyclovir
- Valacyclovir
- Palivizumab
- Enfuvirtide (T-20)
- Maraviroc (MVC)
- Efavirenz / Emtricitabine / Tenofovir (EFV/FTC/TDF)
- Darunavir (DRV)
- Raltegravir (RAL)

# Chapter 5 – Neuro

- **Benzocaine**
- **Lidocaine**
- **Meclizine**
- **Acetaminophen PM**
- Eszopiclone
- Zolpidem
- Ramelteon
- Citalopram
- Escitalopram
- Sertraline
- Fluoxetine
- Paroxetine
- Duloxetine
- Venlafaxine
- Amitriptyline
- Isocarboxazid
- Bupropion
- Varenicline
- Alprazolam
- Midazolam
- Clonazepam
- Lorazepam
- Dexmethylphenidate
- Methylphenidate
- Atomoxetine
- Lithium
- Chlorpromazine
- Haloperidol
- Risperidone
- Quetiapine
- Carbamazepine
- Divalproex
- Phenytoin
- Gabapentin
- Pregabalin
- Levodopa / carbidopa
- Selegiline
- Memantine
- Donepezil
- Scopolamine

## Chapter 6 – Cardio

**Omega-3-Acid E.E.**
**Niacin**
**Aspirin (Low Dose)**
Mannitol
Furosemide
Hydrochlorothiazide
HCTZ / Triamterene
Spironolactone
Potassium Chloride
Doxazosin
Clonidine
Propranolol
Atenolol
Metoprolol Succinate
Metoprolol Tartrate
Carvedilol
Enalapril
Lisinopril
Losartan
Olmesartan
Valsartan
Diltiazem
Verapamil
Amlodipine
Nifedipine
Nitroglycerin
Atorvastatin
Rosuvastatin
Fenofibrate
Heparin
Enoxaparin
Warfarin
Dabigatran
Clopidogrel
Digoxin
Atropine

## Chapter 7 Endocrine and Misc.

**Regular Insulin**
**NPH Insulin**
**Levonorgestrel**
Metformin
Sitagliptin
Glipizide
Glyburide
Glucagon
Insulin lispro
Insulin glargine
Levothyroxine
Propylthiouracil
Testosterone
Ethinyl Estradiol / Norethindrone / Ferrous fumarate (Loestrin 24 Fe)
Ethinyl Estradiol / Norgestimate (Tri-Sprintec)
Ethinyl Estradiol / Etonogestrel (NuvaRing)
Ethinyl Estradiol / Norelgestromin (OrthoEvra)
Oxybutynin
Solifenacin
Tolterodine
Bethanechol
Sildenafil
Tadalafil
Alfuzosin
Tamsulosin
Dutasteride
Finasteride

(Oral contraceptives brand names in parenthesis for clarity)

## THE OTC SCAVENGER HUNT

I learned drug names by working in a pharmacy. I recommend you start learning them with this lab activity. Medications you have held will be easier to memorize. If you are in the car driving and listening to the audio version of this book, obviously, just keep going, but I encourage you to try this activity when you have a chance.

**Picture finding the drugs in an alphabetical list. It's not very conducive to memorization.**

Acetaminophen
Acetaminophen PM
ASA/APAP/Caffeine
Aspirin (Low Dose)
Aspirin (Regular)
Benzocaine
Bismuth subsalicylate
Butenafine
Calcium carbonate
Cetirizine
Diphenhydramine
Docosanol
Docusate sodium
Esomeprazole
Famotidine
Guaifenesin/DM
Ibuprofen
Influenza vaccine
NPH insulin
Regular insulin
Levonorgestrel
Lidocaine
Loperamide
Loratadine
Loratadine-D
Magnesium hydroxide
Meclizine
Naproxen
Neomycin / Polymyxin B / Bacitracin
Niacin
Omega-3 E.E.
Omeprazole
Oxymetazoline
Phenylephrine
Polyethylene gly.
Pseudoephedrine
Ranitidine
Triamcinolone

**Now picture (or actually find) this list sorted by pathophysiologic class. This is how pharmacies sort over-the-counter (OTC) drugs for placement on drug store shelves.**

**Gastrointestinal**

Calcium carbonate
Magnesium hydroxide
Famotidine
Ranitidine
Esomeprazole
Omeprazole
Bismuth subsalicylate
Loperamide
Docusate sodium
Polyethylene gly.

**Musculoskeletal**

Aspirin (Regular)
Ibuprofen
Naproxen
Acetaminophen
Acetaminophen / Aspirin / Caffeine

**Respiratory**

Diphenhydramine
Cetirizine
Loratadine
Loratadine-D
Pseudoephedrine
Phenylephrine
Oxymetazoline
Triamcinolone
Guaifenesin / DM

**Immune**

Neomycin /
Polymyxin B /
Bacitracin
Butenafine
Docosanol
Influenza vaccine

**Neuro**

Benzocaine
Lidocaine
Meclizine
Acetaminophen PM

**Cardio**

Omega-3-Fatty E.E.
Niacin
Aspirin (Low Dose)

**Endocrine**

Regular insulin
NPH insulin
Levonorgestrel

You will find the second list often grouped together in the pharmacy aisles. The same is true in your brain. It remembers drugs in related groups, not in strict alphabetical order. In each chapter, I will build on this concept providing:

- Drug class summaries
- Brand / generic pronunciations
- Mnemonics created by students and instructors
- Two quizzes (one easier / one harder)
- A comprehensive chapter mnemonic

# CHAPTER 1 GASTROINTESTINAL

## I. Peptic ulcer disease

Clinicians diagnose peptic ulcer disease (PUD) when an ulceration of the peptic or digestive tract occurs. Acid is an aggressive factor in the stomach that, if reduced, allows an ulcer to heal. In this chapter we'll focus on antacids, histamine$_2$ receptor antagonists (also known as $H_2$ blockers), and proton pump inhibitors. Note: antibiotics eradicate *Helicobacter pylori,* the organism responsible for the ulcers, but we will tackle those in chapter 4.

### Antacids

Antacids, or literally, anti-acids, can contain the elements **calcium** and **magnesium** to raise the stomach's pH. The *pH scale* is like a one-foot long ruler with 14 lines instead of 12. A "0" sits left for the most acidic compounds and a "14" sits right for the most basic or alkaline ones. 7, which is in the middle, is neutral. Stomach acid has a pH of 2; milk is 6; neutral is 7; and blood is 7.35. Why give dairy milk, which is slightly acidic, to calm an acidic stomach? Because 6 is more basic (alkaline) than 2.

Besides acting as an antacid, **calcium carbonate (Tums)** can supplement calcium in a diet and **magnesium hydroxide (Milk of Magnesia)** works as a laxative. Unfortunately, both antacids can **chelate** (bind with) antibiotics like **doxycycline (Doryx)** and **ciprofloxacin (Cipro).**

**Calcium carbonate (Tums, Children's Pepto)**
*CAL-see-um CAR-bow-nate (TUMS)*

> Students associate the brand name Tums with the word tummy to remember it's an antacid. **Calcium carbonate (Children's Pepto)** and **bismuth subsalicylate (Adult Pepto-Bismol)** have different ingredients and should not be interchanged.

**Magnesium Hydroxide (Milk of Magnesia)**
*mag-KNEES-e-um high-DROCKS-ide (MILK mag-KNEE-shuh)*

> **Milk of Magnesia** looks like dairy milk, which can work similarly to an antacid in calming an acidic stomach. Put together the "milky texture" and the diarrhea of lactose intolerance to remember its laxative effect.

## HISTAMINE$_2$ RECEPTOR ANTAGONISTS ($H_2$RAS)

The term $H_2$ blocker (more formally, $H_2$ receptor antagonist) stands for histamine receptor type two ($H_2$) blocker. Histamine$_2$ causes the formation of acid, so blocking its receptors reduces the production of acid. You will notice **ranitidine (Zantac)** and **famotidine (Pepcid)** both end in "tidine." Related drugs often have related stems in their names.

When someone says, "I need an antihistamine," they are generally looking for relief from allergic symptoms like sneezing, runny nose, watery eyes, etc. Those allergy antihistamines affect histamine one ($H_1$) receptors. We will cover those in the respiratory chapter.

**Famotidine (Pepcid)**
*Fa–MOE–ti–dean (PEP-sid)*

**Famotidine** has the "-tidine" stem indicating it's an $H_2$ blocker. **Pepcid** contains "pep" from "peptic" and "pepsin" that relates to digestion, and "cid" from "acid." One student mentioned Pepsi, the soda, has a pH of 2.4, which is very acidic. She said it always gave her heartburn, so that's how she remembered **Pepcid**. Also, in soda pop, the "p-o-p" and **Pepcid's** "p-e-p" are very similar.

**Ranitidine (Zantac)**
*ra-NI-ti-dean (ZAN-tack)*

One student said **ranitidine's** "-tidine" looks like "to dine," the time patients might experience gastroesophageal reflux disease (GERD). **Zantac** looks like a "2" with "antac" after it, so it's an $H_2$ blocker, which works as an ant-agonist to ac-id.

## PROTON PUMP INHIBITORS (PPIS)

Proton pump inhibitors (PPIs) block a pump that introduces protons (which are acidic) into the stomach, thus making it less acidic. Prescribers combine PPIs with antibiotics for ulcer triple therapy to kill *H. pylori*.

While **esomeprazole (Nexium)** and **omeprazole (Prilosec)** have the same ending "prazole," notice the only thing that separates **omeprazole** and **esomeprazole** is an "es." Many drugs have chemical structures that have mirror images called enantiomers. Instead of calling them right- and left-handed, we call them "R" and "S" from the Latin words *rectus* (right) and *sinister* (left).

In this case, the "S" form is more active biologically. Putting an "s" in front of omeprazole would make "someprazole," which would be pronounced "some-prazole." Instead, the prefix "es" allows for a separation between the "S" sound and the compound name, as a chemist would pronounce it.

**Esomeprazole (Nexium)**
*es-oh-MEP-rah-zole (NECKS-see-um)*

> The "prazole" ending means PPI and the "e-s" means "S" for sinister or left-handed. The manufacturer released **Nexium** after **Prilosec** as the "next" PPI drug.

**Omeprazole (Prilosec)**
*oh-MEP-rah-zole (PRY-low-sec)*

> The "prazole" ending in **omeprazole** indicates proton pump inhibitor (PPI). We remember **Prilosec** by the "Pr" for hydrogen protons (protons, or hydrogen ions, are what make an acid acidic), the "lo" for low, and the "sec" for secretion of those protons. Or, the "o" in Prilosec looks like a zero, and **Pril-"O"-sec** provides zero heartburn.

## II. DIARRHEA, CONSTIPATION, AND EMESIS

Diarrhea can lead to dehydration and sometimes we need to intervene and use over-the-counter medications like **bismuth subsalicylate (Pepto-Bismol)** or **loperamide (Imodium)**.

Bismuth's "**subsalicylate**" is similar to aspirin (acetyl**salicylic** acid), and is dangerous to young children. **Bismuth subsalicylate (Pepto Bismol)** is not appropriate for children because of the risk of **Reye's syndrome**, a condition involving brain and liver damage that can occur in children with chicken pox or influenza who take **salicylates**.

Opioids like **morphine (Kadian)** decrease gastrointestinal tract motility, causing constipation. Calcium channel blockers like **verapamil (Calan)** block calcium from getting to the bowel's smooth muscle. Prescribers frequently give a stool softener like **docusate sodium (Colace)** and a stimulant laxative for constipation caused by these medications, or due to other causes.

Emesis or vomiting is an effective natural body response to ingested toxins. However, with cancer chemotherapy, we want to prevent chemotherapy-induced nausea and vomiting (CINV) with drugs like **ondansetron (Zofran)**.

Manufacturers specially formulate these meds because nausea patients may vomit oral meds. For example, **ondansetron (Zofran)** comes as an orally disintegrating tablet (ODT) that patients take without water, and **promethazine (Phenergan)** as a rectal suppository.

## ANTIDIARRHEALS

**Bismuth Subsalicylate (Pepto-Bismol)**
*BIZ-muth sub-sal-IS-uh-late (pep-TOE BIZ-mol)*

> The "b" in **bismuth subsalicylate** reminds students of the black tongue and black stool that some patients experience as side effects. (Note: This discoloration is harmless.) **Pepto** looks like peptic, which has to do with digestion.

**Loperamide (Imodium)**
*Low-PER-uh-mide (eh-MOE-dee-um)*

The "lo" for slow and "per" for peristalsis is how a student remembered **loperamide's** function. **Imodium** is like the word "immobile" in that it slows down the bowel.

## CONSTIPATION – STOOL SOFTENER

**Docusate Sodium (Colace)**
*DOCK-you-sate SEWED-e-um (CO-lace)*

**Docusate sodium** softens the stool. Patients use it with opioids like **morphine (Kadian)**. **Docusate** and "penetrate" rhyme, and **docusate sodium** works by helping water penetrate into the bowel. The brand **Colace** improves the colon's pace.

## CONSTIPATION – OSMOTIC

**Polyethylene Glycol (PEG) 3350 (MiraLax)**
*pa-Lee-ETH-ill-een GLY-call (MIR-uh-lacks)*

I remember **polyethylene glycol** because I have triplets "**poly**" calling "**col**" for me, "Daaaaaad! Can you wipe me?" **MiraLax** is the Miracle Laxative because it's a miracle how good you feel after taking it. By prescription, **Go-Lytely** is a 4-liter plastic bottle of **polyethylene glycol** used for colonoscopy examination preparation. There is nothing "lightly" about it.

### ANTIEMETIC - SEROTONIN 5-HT3 RECEPTOR ANTAGONIST

**Ondansetron (Zofran, Zofran ODT)**
*on-DAN-se-tron (ZO-fran)*

The "setron" suffix will help you remember **ondansetron** is a serotonin 5-HT3 receptor antagonist for preventing emesis. "O-D-T" stands for orally disintegrating tablet. It is a useful dosage form because it dissolves on the top of the tongue and requires no additional liquid. If you are good at word scrambles, **ondansetron** has every letter but the "i" in serotonin, a neurotransmitter, the majority of which is located in the GI tract.

### ANTIEMETIC - PHENOTHIAZINE

**Promethazine (Phenergan)**
*pro-METH-uh-zeen (FEN-er-gan)*

**Promethazine** is an antihistamine sometimes used in liquid form with codeine. It also reduces nausea. In addition to oral, IM, and IV forms, **promethazine** comes in a rectal suppository form if a patient can't take anything by mouth (po).

## III. GASTROINTESTINAL AUTOIMMUNE DISORDERS

Autoimmune diseases like ulcerative colitis (UC) occur when the body's immune system inappropriately attacks an area (or areas) of the body. Symptoms of UC include ulcerations and inflammation (-itis) in the colon. **Infliximab**

**(Remicade)** blocks the tumor necrosis factor, alpha (TNF-alpha), to treat this disease.

## ULCERATIVE COLITIS

**Infliximab (Remicade)**
*in-FLIX-eh-mab (REM-eh-cade)*

> **Infliximab** is a biologic agent, a genetically engineered protein. **Infliximab's** generic name should be broken up as inf + li +xi + mab. The "inf" is a prefix that simply separates it from other similar drugs. The "li" stands for immunomodulator (the target). The "xi" stands for chimeric (the source, e.g., combining genetic material from a mouse, with genetic material from a human). The "x" might also refer to the Greek letter "chi," which looks like an x. The "mab" stands for monoclonal antibody. Conditions like ulcerative colitis can go into remission. **Remicade** is a "remission aide."

## GASTROINTESTINAL DRUG QUIZ (LEVEL 1)

Classify these drugs by placing the corresponding drug class letter next to each medication. Try to underline the stems before you start and think about the brand name and function of each drug.

1. Calcium carbonate (Tums)
2. Ranitidine (Zantac)
3. Docusate sodium (Colace)
4. Famotidine (Pepcid)
5. Esomeprazole (Nexium)
6. Loperamide (Imodium)
7. Magnesium hydroxide (Milk of Magnesia)
8. Infliximab (Remicade)
9. Omeprazole (Prilosec)
10. Ondansetron (Zofran)

**Gastrointestinal drug classes:**

A. Antacid
B. Anti-diarrheal
C. Anti-nausea
D. Constipation
E. $H_2$ blocker
F. Proton pump inhibitor
G. Ulcerative colitis

## GASTROINTESTINAL DRUG QUIZ (LEVEL 2)

Classify these drugs by placing the corresponding drug class letter next to each medication. Try to underline the stems before you start and remember the brand name and function of each drug.

1. Bismuth Subsalicylate
2. Esomeprazole
3. Omeprazole
4. Polyethylene glycol
5. Promethazine
6. Calcium carbonate
7. Famotidine
8. Docusate sodium
9. Loperamide
10. Ranitidine

**Gastrointestinal drug classes:**

A. Antacid
B. Anti-diarrheal
C. Anti-nausea
D. Constipation
E. $H_2$ blocker
F. Proton pump inhibitor
G. Ulcerative colitis

## GI: MEMORIZING THE CHAPTER

Now that you have had a chance to get to know each of the drugs in the list individually, you are ready to memorize the first thirteen medications in order. I have included the connections I made to assist you in memorizing them, but if you have created better ones, use them. Some students copy and paste the GI drug table on the front of a 4 x 6 card and the "brain shorthand" explanation on the back. Remember, **bolded** drugs are OTC.

| **Calcium carbonate** | **Bismuth subsalicylate** | Infliximab |
|---|---|---|
| **Magnesium hydroxide** | **Loperamide** | |
| **Famotidine** | **Docusate sodium** | |
| **Ranitidine** | **Polyethylene glycol** | |
| **Esomeprazole** | Ondansetron | |
| **Omeprazole** | Promethazine | |

## "G" GASTROINTESTINAL

Broadly speaking, there are 13 drugs in the gastrointestinal section. The first six are acid reducers, the next two are antidiarrheals, the next two are laxatives, the next two are antiemetics, and final one is for ulcerative colitis.

Picture in your mind where in the human body these work. Start with the six acid reducers in the stomach, move to the two laxatives and antidiarrheals that work in the intestines / colon, go back up to the antiemetics to prevent vomiting from the mouth, and back down to the colon to the ulcerative colitis medication. Why this up and down and up and down? It's easiest to start at the stomach and work down to the intestines with OTC drugs, then work top to bottom from the mouth to the intestines with RX drugs.

I ordered the first three pairs of acid reducers from the fastest to slowest working: antacids take a few minutes, $H_2$Blockers take about half an hour, and proton pump inhibitors (PPIs) can take as long as a day. "A" for antacid is first. $H_2$Blocker has "2" for the second group. PPI has three letters for third group. The antacids **calcium carbonate** and **magnesium hydroxide** are both in the same column on the periodic table of elements in alphabetical order. Two $H_2$ Blockers: **famotidine** and **ranitidine** follow. Two PPIs **esomeprazole** and **omeprazole** follow them. Often diarrhea follows an upset stomach, so we treat with **bismuth subsalicylate** and **loperamide**. Use "L" from loperamide to get to "L" for two laxatives: **docusate sodium** and **polyethylene glycol**. Use the "p-o" from polyethylene reversed as "o-p" for **ondansetron** and **promethazine**. You can rectally administer promethazine next to colon to get to the ulcerative colitis medication – **infliximab**.

Can you recite in order: **calcium carbonate, magnesium hydroxide, famotidine, ranitidine, esomeprazole, omeprazole, bismuth subsalicylate, loperamide, docusate sodium, polyethylene glycol, ondansetron, promethazine, infliximab?**

Do you recognize the stems –tidine, -prazole, -sal-, -setron, -liximab and the therapeutic classes they represent?

Once you've answered these two questions for this chapter and the following chapters, you are ready to move on.

# CHAPTER 2 MUSCULOSKELETAL

## I. NSAIDs AND PAIN

Non-steroidal anti-inflammatory drugs (NSAIDs) (pronounced *EN-SAIDs*) contrast with the steroidal medications used to treat inflammation. The NSAIDs **aspirin (Ecotrin)** and **ibuprofen (Advil, Motrin)** [taken up to four times daily] and **naproxen (Aleve)** [taken twice daily] are available over-the-counter and are for sporadic or mild pain. **Meloxicam (Mobic)** [once daily] isn't OTC and has the longest half-life. *Analgesics* relieve pain. *Antipyretics* reduce fever. NSAIDs are both.

NSAIDs like **ibuprofen** can close an arterial shunt (*patent ductus arteriosus*) in a preemie. Our newborn daughter had this condition. I watched a YouTube video on the surgery to close the shunt and it took only five minutes, but surgery on any NICU neonate runs a great risk. We were thankful a simple NSAID closed the shunt.

A regular pharmacy question is when to use **acetaminophen (Tylenol)** and when to use an **NSAID**. If the patient has inflammation, prescribers prefer NSAIDs, as analgesics like **acetaminophen** will not help. However, if the patient has pain or fever, then either is appropriate. Generally, NSAIDs are not used for pregnant patients, but **acetaminophen** (**Tylenol**) can be.

**Excedrin Migraine** contains three drugs: **aspirin** for pain and inflammation, **acetaminophen** for pain, and **caffeine** as a potent vasoconstrictor. It narrows swollen brain vessels.

Common NSAIDs block both cyclooxygenase (COX) -1 and -2 to minimize inflammatory processes. COX-1 blockade reduces inflammation in the body and, unfortunately, the body's natural protection of the stomach lining. Note that a selective COX-2 inhibitor, like **celecoxib (Celebrex)** does not block the protective effect of COX-1 against stomach ulcers and this is why it's supposed to be a better choice.

## OTC Analgesics – NSAIDs

**Aspirin [ASA] (Ecotrin)**
*AS-per-in (ECK-oh-trin)*

The acronym for aspirin, "ASA," comes from the chemical name: AcetylSalicylicAcid. **Ecotrin** is enteric-coated aspirin.

**Ibuprofen (Advil, Motrin)**
*eye-byou-PRO-fin (ADD-vil, MO-trin)*

Many students try to say that **ibuprofen** and **naproxen** both end with "en." However, many drugs end with "e-n," so that won't help in a large multiple-choice exam. A better mnemonic is to notice that "-profen" from **ibuprofen** is a recognized stem and differs from "proxen" in **naproxen** by only one letter.

**Naproxen (Aleve)**
*nap-ROCKS-in (uh-LEAVE)*

While **naproxen** has no formal stem, a student came up with a mnemonic for the brand name: **Aleve will** alleviate pain from strains and sprains.

## OTC Analgesic – Non-narcotic

**Acetaminophen [APAP] (Tylenol)**
*uh-seat-uh-MIN-no-fin (TIE-len-all)*

The brand **Tylenol**, generic **acetaminophen**, and acronym **APAP** all come from the chemical name:

N-acetyl-para-amino-phenol (**Tylenol**)
N-acetyl-para-amino-phenol (**Acetaminophen**)
N-Acetyl-Para-Amino-Phenol (**APAP**)

## OTC Migraine – NSAID / Non-narcotic analgesic

**Aspirin / Acetaminophen / Caffeine (Excedrin Migraine)**
*AS-per-in / uh-seat-uh-MIN-oh-fin / KAF-feen (ecks-SAID-rin)*

Most students remember **Excedrin Migraine** by the rationale for the combination of **ASA / APAP / caffeine**: inflammation / analgesia / vasoconstriction.

## RX Analgesics – NSAIDs

**Meloxicam (Mobic)**
*mel-OX-eh-kam (MO-bik)*

The "-icam" suffix in the generic name **meloxicam** lets you know it's an NSAID. A student used the "bic" in **Mobic** to remember it treats "big" swelling.

## RX Analgesics – NSAIDs – COX-2 inhibitor

**Celecoxib (Celebrex)**
*sell-eh-COCKS-ib (SELL-eh-breks)*

The "–coxib" suffix lets you know celecoxib is a selective COX-2 inhibitor. The commercials for **Celebrex** talk about celebrating relief from inflammatory conditions. Regular NSAIDs like **ibuprofen** and **naproxen** inhibit both COX-1 and COX-2, causing the stomach distress so commonly caused by drugs in the NSAID class. **Celebrex** causes less GI irritation due to its lack of COX-1 inhibition.

# II. Opioids and narcotics

**Opioids** relieve pain, but have addiction potential. The DEA (Drug Enforcement Agency) categorizes the addictive potential of medications using a drug scheduling system.

*Schedule I* drugs are illegal substances and have no medical value, such as **heroin**.

*Schedule II* drugs are potentially addicting, such as the individual drugs **fentanyl (Duragesic, Sublimaze) and morphine (Kadian),** and the combination products **hydrocodone / APAP (Vicodin)** and **oxycodone / APAP (Percocet)**.

*Schedule III* drugs are less addicting and include **acetaminophen / codeine (Tylenol with codeine)**.

*Schedule IV* drugs include some sedative-hypnotics (sleeping pills) such as **zolpidem (Ambien)** and mixed opioid analgesics like **tramadol (Ultram)**.

*Schedule V* drugs are often cough medicines like **guaifenesin / codeine (Cheratussin AC)** that include codeine, but not **codeine** as a drug alone.

An opioid side effect is pinpoint pupils called miosis. A student remembered this by noticing that the words "opioid" and "miosis" both have two little dots over the two i's that look like pinpoint pupils.

## Opioid Analgesics – Schedule II

**Morphine (Kadian, MS Contin)**
*MORE-feen (KAY-dee-en, EM-ES KON-tin)*

The generic name **morphine** comes from the ancient Greek god of dreams, Morpheus. **Kadian** might come from cir**cadian** (the twenty-four-hour cycle) because **Kadian** is an extended-release morphine formulation. **MS Contin** stands for morphine sulfate **contin**uous release.

**Fentanyl (Duragesic, Sublimaze)**
*FEN-ta-nil (dur-uh-GEE-zic, SUB-leh-maze)*

**Fentanyl** is troubling because it's dosed in micrograms, not milligrams. When it was time to give our three-month-old preemie **fentanyl** after her pyloric valve stenosis surgery in the Neonatal Intensive Care Unit (NICU), I made sure to check the calculated dose. **Duragesic** is a long **dura**tion anal**gesic** and comes in a patch that provides relief for 72 hours. **Sublimaze** is an injectable form of **fentanyl**.

**Hydrocodone / Acetaminophen (Vicodin)**
*high-droe-CO-done / uh-seat-uh-MIN-no-fin (VIE-co-din)*

**Hydrocodone** and **oxycodone** differ slightly in chemical structure. **Oxycodone** has one more oxygen than **hydrocodone. Oxycodone/APAP** gets its name from the first three letters of *oxy*gen. **Hydrocodone/APAP** has only one hydrogen and gets its name from the first five letters of *hydro*gen. The "c-o-d-i-n" in **Vicodin** looks like **codeine,** just drop two e's from codeine, so you can remember they are related.

**Oxycodone/Acetaminophen (Percocet)**
*ox-e-CO-done / uh-seat-uh-MIN-no-fin (PER-coe-set)*

The "cet" in **Percocet** comes from acetaminophen. Some students use that "codone" and "codeine" look a little alike to remember the similarity, but this is not a true stem and **Percocet** does not contain codeine. Drug companies add **acetaminophen** as a mild analgesic.

## Opioid Analgesics – Schedule III

**Acetaminophen/Codeine (Tylenol/Codeine, Tylenol #3)**
*uh-seat-uh-MIN-no-fin with CO-dean*
*(TIE-len-all with CO-dean, TIE-len-all NUM-ber three)*

I am not sure why, but when you say the generic name of **Vicodin,** you say "**hydrocodone with acetaminophen,**" but when you talk about **codeine,** the order is reversed. It's phrased as "**acetaminophen with codeine" (Tylenol w/codeine).**

Students seem to remember both because of the reverse / opposite order. The "#3" in **Tylenol #3** refers to the amount of codeine in combination. For example:

**Tylenol #2** 15 mg codeine / 300 mg acetaminophen
**Tylenol #3** 30 mg codeine / 300 mg acetaminophen
**Tylenol #4** 60 mg codeine / 300 mg acetaminophen

## Mixed-opioid receptor analgesic – Schedule IV

**Tramadol (Ultram)**
*TRA-muh-doll (ULL-tram)*

**Tramadol** only weakly affects opioid receptors. For this reason, the DEA did not classify **tramadol** as a controlled substance until 2014. The "-adol" stem indicates that it's a mixed opioid analgesic. Many students have thought of "tram wreck" and "train wreck" as a way to remember that **Ultram** is used for pain.

## Opioid antagonist

**Naloxone (Narcan)**
*nah-LOX-own (NAR-can)*

**Naloxone** is an opioid receptor antagonist used in opioid overdose situations. Often it's in the L-E-A-N acronym of emergency medicines **lidocaine**, **epinephrine**, **atropine**, and **naloxone**. The "nal-" stem in **naloxone** indicates it's an opioid receptor antagonist, but the brand name **Narcan** also hints at a narcotic antagonist.

## III. HEADACHES AND MIGRAINES

Common OTC drugs for headache and migraine include the NSAIDs like **aspirin (Ecotrin), ibuprofen (Advil, Motrin), naproxen (Aleve),** and the combination **ASA / APAP / Caffeine (Excedrin Migraine)** reviewed earlier.

Sometimes severe migraines require drugs in the 5-$HT_1$ receptor agonist class, such as **ele<u>triptan</u> (Relpax)** and **suma<u>triptan</u> (Imitrex)** that work by activating receptors that reduce the swelling associated with migraines. We call them **triptans** because those two syllables are easier to say than "5-hydroxytryptamine receptor agonists" is.

It's important to contrast the terms **agonist** and **antagonist** as opposites. An agonist is a drug that activates a receptor while an antagonist blocks the receptor. There is no reason to *produce* migraines, but a 5-$HT_1$ receptor *antagonist* would likely cause a headache, whereas the corresponding agonist alleviates headaches. One student thought of agonists as when she first stepped on an elliptical exercise machine and it awakened the screen. She thought of antagonist as her own excuses why she could not work out. Another metaphor you will often see is that of dating. There are some funny videos on YouTube about agonism versus antagonism. The agonist is usually someone who wants to get a date and the antagonist tries to block the suitor.

### 5-$HT_1$ RECEPTOR AGONIST

**Ele<u>triptan</u> (Relpax)**
*el-eh-TRIP-tan (IM-eh-treks)*

Remember **eletriptan** from its suffix "–triptan" so you can recognize the many other medications in the triptan class. The brand name **Relpax** combines "pax," the Latin word for "peace" and "Rel," for "relief." I often confused whether this was an agonist or antagonist, but use the "agony" of a migraine to remember triptans as agonists.

**Suma<u>triptan</u> (Imitrex)**
*Sue-ma-TRIP-tan (IM-eh-treks)*

Remember **sumatriptan** from its suffix "-triptan." Students say it "trips up" a headache. Also, the "i-m" in the brand name **Imitrex** reminds you there is an intramuscular (IM) form for patients who have such a severe migraine that they can't take anything by mouth.

## IV. DMARDs and Rheumatoid Arthritis

**DMARD** stands for <u>D</u>isease-<u>M</u>odifying <u>A</u>nti-<u>R</u>heumatic <u>D</u>rug, which means it works against *rheumatoid arthritis*, an autoimmune disorder. These drugs reduce the progression of the disease as opposed to the treatment of *osteoarthritis*, a condition in which the body has worn down and the joints are inflamed.

Both conditions respond to NSAID anti-inflammatories such as **ibuprofen (Advil, Motrin)** or **aspirin (Ecotrin)**. Additionally, glucocorticoids, such as **prednisone (Deltasone)**, can further help reduce inflammation in the joints.

Prescribers use special immune-suppressing drugs called DMARDs like **methotrexate (Rheumatrex)**, **abatacept (Orencia)** and **etanercept (Enbrel)** for rheumatoid arthritis.

## DMARDs

**Methotrexate (Rheumatrex)**
*meth-oh-TREKS-ate (ROOM-uh-treks)*

One student came up with "Meth o T-Rex ate the rheumatic inflammate." The "rheuma" in the brand name **Rheumatrex** reminds you it relieves rheumatoid arthritis.

**Abatacept (Orencia)**
*uh-BAT-uh-sep" (or-EN-see-uh)*

**Abatacept** and **etanercept** have complex stems. The "-ta-" infix in **abatacept** means it's going after T-cell receptors and the suffix "–cept" means that it's a receptor molecule, either native or modified.

**Etanercept (Enbrel)**
*eh-ta-NER-sept (EN-brell)*

Learn **etanercept** and **abatacept** from the suffix "-cept" with the added sub-stem "-tacept" or "-nercept" respectively. The "-ner-" infix in **etanercept** point out that it goes after tumor necrosis factor receptors.

# V. Osteoporosis

Don't confuse *osteoarthritis*, a joint disease, with *osteoporosis*, a thinning in bone tissue density. Drugs for osteoarthritis include the NSAIDs. Drugs for osteoporosis build bone back up. Humans' bones grow slowly. Therefore, drugs like **alendronate (Fosamax)** can be given weekly, and **ibandronate (Boniva)** can be given monthly. A special precaution for patients using **alendronate** is to not lie down for 30 minutes after taking the medication; for **ibandronate**, it's 60 minutes.

### Bisphosphonates

**Alendronate (Fosamax)**
*uh-LEN-dro-nate (FA-seh-max)*

Memorize **alendronate** as a calcium metabolism regulator from its "-dronate" stem. Students like to remember that "drone" rhymes with bone. Many students have said that the brand name **Fosamax** looks like "fossil."

**Ibandronate (Boniva)**
*eh-BAND-row-nate" (bo-KNEE-vuh)*

Again, the "–dronate" suffix should be your key to the drug class, but having the first three letters of bone in the brand name **Boniva** helps. The generic **ibandronate** contains all the letters in **Boniva** except the "v."

## VI. Muscle relaxants

The DEA doesn't schedule **cyclobenzaprine (Flexeril)**, but does schedule **diazepam (Valium)** as C-IV. Both drugs provide muscle relaxation and relief from muscle spasms.

**Cyclobenzaprine (Flexeril)**
*sigh-clo-BENDS-uh-preen (FLEX-er-ill)*

**Cyclobenzaprine** helps you get bending again. **Flexeril** improves flexibility.

**Diazepam (Valium)**
*dye-AY-zeh-pam" (VAL-e-um)*

**Diazepam** and benzo**diazep**ine, diazepam's drug class, have similar letters. **Val**erian root is an herbal remedy for anxiety; some think of **Valium** as relaxing both anxiety and muscles in the same way.

# VII. Gout

**Gout** is an acute inflammatory arthritis we treat acutely (right away) with an NSAID like **ibuprofen** (**Advil, Motrin**) We can treat also treat gout prophylactically by reducing uric acid, a major component in the crystals that cause the gouty pain. Drugs that alter uric acid levels include **allopurinol (Zyloprim)** and **febuxostat (Uloric)**.

## Uric Acid Reducers

**Allopurinol (Zyloprim)**
*a-loe-PURE-in-all (ZY-low-prim)*

Within **allopurinol**, you can see "uri," which corresponds to the uric acid the medication reduces. You can also remember this is an anti-arthritic by thinking of the joints as becoming "all-pure-and-all."

**Febuxostat (Uloric)**
*fe-BUCKS-oh-stat (YOU-lore-ick)*

The "-xostat" stem in **febuxostat** indicates a xanthine oxidase inhibitor that prevents uric acid from forming. The "x" and "o" in the generic name match xanthine oxidase. The brand name **Uloric** looks like "U" "lower" "uric acid."

# MUSCULOSKELETAL DRUG QUIZ (LEVEL 1)

Classify these drugs by placing the corresponding drug class letter next to each medication. Try to underline the stems before you start and think about the brand name and function of each drug.

1. Acetaminophen (Tylenol)
2. Alendronate (Fosamax)
3. ASA/APAP/Caffeine (Excedrin)
4. Febuxostat (Uloric)
5. Etanercept (Enbrel)
6. Fentanyl (Duragesic)
7. Hydrocodone / APAP (Vicodin)
8. Ibuprofen (Advil, Motrin)
9. Celecoxib (Celebrex)
10. Sumatriptan (Imitrex)

**Musculoskeletal Drug Classes:**

A. 5-$HT_1$ receptor agonist
B. Anti-gout
C. Bisphosphonate
D. DMARD
E. Non-narcotic analgesic – single
F. Non-narcotic analgesic combo - headache
G. NSAID not COX-2 selective
H. NSAID COX-2 selective
I. Opioid analgesic

# MUSCULOSKELETAL DRUG QUIZ (LEVEL 2)

Classify these drugs by placing the corresponding drug class letter next to each medication. Try to underline the stems before you start and remember the brand name and function of each drug.

1. Methotrexate
2. Morphine
3. Abatacept
4. Naproxen
5. Febuxostat
6. Ibandronate
7. Allopurinol
8. Aspirin
9. Alendronate
10. APAP/Codeine

**Musculoskeletal Drug Classes:**

A. $5\text{-HT}_1$ receptor agonist
B. Anti-gout
C. Bisphosphonate
D. DMARD
E. Non-narcotic analgesic - single
F. Non-narcotic analgesic combo - headache
G. NSAID not COX-2 selective
H. NSAID COX-2 selective
I. Opioid analgesic

# MUSCULOSKELETAL: MEMORIZING THE CHAPTER

| **Aspirin** | Morphine | Methotrexate |
|---|---|---|
| **Ibuprofen** | Fentanyl | Abatacept |
| **Naproxen** | Hydrocodone/APAP | Etanercept |
| **Acetaminophen** | Oxycodone/APAP | Alendronate |
| **ASA/APAP/Caffeine** | APAP/Codeine | Ibandronate |
| Meloxicam | Tramadol | Cyclobenzaprine |
| Celecoxib | Naloxone | Diazepam |
| | Eletriptan | Allopurinol |
| | Sumatriptan | Febuxostat |

## "M" MUSCULOSKELETAL

Three OTC NSAIDs, **aspirin, ibuprofen, naproxen**, and then **acetaminophen,** a non-narcotic analgesic. Combine **aspirin / acetaminophen** and **caffeine** to make **Excedrin Migraine.** Use migraine (M) and caffeine (C) to connect **meloxicam** (M) and **celecoxib** (C). Group opioids by DEA class. Start with DEA Schedule II **morphine**, the original agent; then **fentanyl**, **hydrocodone / APAP,** and **oxycodone / APAP**. Next comes schedule III, **acetaminophen with codeine**; to schedule IV **tramadol**; then to the opioid *narc*otic *antagonist* **naloxone (Narcan);** to migraine agony *agonists* for migraine headache with **eletriptan** and **sumatriptan**; down to rheumatoid arthritis inside the joints with the DMARD **methotrexate**, a non-biologic; to two biologic DMARDs **abatacept** to **etanercept**. From joint pain on to the brittle bones with **alendronate** and **ibandronate** for osteoporosis. Then move out to the muscles and muscle relaxers **cyclobenzaprine** and **diazepam** all the way down to the big toe for the gout meds **allopurinol** and **febuxostat**.

# CHAPTER 3 RESPIRATORY

## I. ANTIHISTAMINES AND DECONGESTANTS

We divide **antihistamines** into two generations: first and second. **Diphenhydramine** (**Benadryl**) is a first-generation antihistamine. In addition to its ability to improve allergy symptoms, it usually makes patients sleepy. The second-generation agents cannot pass through the blood-brain barrier and into the central nervous system, which limits the degree of drowsiness patients experience. This second generation includes **cetirizine** (**Zyrtec**) and **loratadine (Claritin**).

My students came up with a useful / helpful / creative visual to remember the difference between histamine-1 and histamine-2 receptors called the antihistamine snowman. They pictured $H_1$ in the snowman's head because he has only one carrot for his nose, and allergies usually happen in the head / nose. They pictured $H_2$ as the snowman's belly because that's where gastroesophageal reflux disease (GERD) and hyperacidity happen.

Nasal decongestants like **pseudoephedrine (Sudafed)** constrict blood vessels in the nose and sinuses, reducing the amount of mucus that is formed, and can be found in combination with antihistamines. Usually you will see the brand or generic name followed by a hyphen "D" for decongestant, like **Claritin-D** to indicate that **pseudoephedrine** is a product ingredient. These are not over-the-counter (OTC), but behind the counter (BTC) because you have to show an ID to purchase them.

## OTC Antihistamine 1st-generation

**Diphenhydramine (Benadryl)**
*dye-fen-HIGH-dra-mean (BEN-uh-drill)*

Many students on YouTube videos and Quizlet notecards mistake **diphenhydramine's** "i-n-e" as a stem because many antihistamines end in "ine." However, so does **morphine** and roughly 20% of all generic names. There is not really a good stem for antihistamines because of this. You can remember the brand name by recognizing **Benadryl** benefits you by drying up your runny nose. Some students also associated the "B" in **Benadryl** with the BBB, the Blood Brain Barrier, which **Benadryl** can pass through. Manufacturers use **diphenhydramine** as the "P-M" in many sleep aids, so associating the "B" in **Benadryl** with bedtime makes sense.

## OTC Antihistamine 2nd-generation

**Cetirizine (Zyrtec)**
*seh-TIE-rah-zine (ZEER-teck)*

Pronounce the "t-i-r" in **cetirizine** as tear, like teardrop, and I think of **cetirizine** protecting me from tearing from eyes affected by allergies.

**Loratadine (Claritin)**
*lore-AT-uh-dean (KLAR-eh-tin)*

The "-atadine" stem (used to be "-tadine") is helpful in distinguishing **loratadine** from the "-tidine" stem in the $H_2$ receptor antagonists **famotidine** and **ranitidine.** The **Claritin** "clear" commercials resonate with the drug's function of clearing one's head from allergies or clear eyes relieved from allergy.

## OTC Antihistamine 2nd-generation / Decongestant

**Loratadine / Pseudoephedrine (Claritin-D)**
*lore-AT-uh-dean / Sue-doe-uh-FED-rin (KLAR-eh-tin dee)*

Because **pseudoephedrine** has 15 letters, manufacturers abbreviate it as "hyphen D" for decongestant. Therefore, adding **loratadine**, an antihistamine, to **pseudoephedrine** a decongestant, helps with both runny and stuffy noses.

## OTC Decongestants

**Pseudoephedrine (Sudafed)**
*Sue-doe-uh-FED-rin (Sue-duh-FED)*

If you take out the second "e" and drop the "rine" from **pseudoephedrine**, you get the pronunciation of **Sudafed**. One student said she was p-h-e-d up with being congested and that's how she remembered it.

**Phenylephrine (NeoSynephrine)**
*FEN-ill-EF-rin (KNEE-oh-sin-EF-rin)*

**Phenylephrine** sounds a bit like **pseudoephedrine** – they are both decongestants. Patients recognize **pseudoephedrine** by the "hyphen D" and **phenylephrine** by the "PE" abbreviation. **Phenylephrine** is not as strong as **pseudoephedrine**

and that's one reason it's available OTC, while **pseudoephedrine** is not.

**Oxymetazoline (Afrin)**
*ox-EE-meh-taz-oh-lin (AF-rin)*

**Oxymetazoline** takes us from physically behind-the-counter (BTC) oral **pseudoephedrine** to over-the-counter oral **phenylephrine** to the intranasal decongestant **oxymetazoline**. The "**Afrin**" brand name sounds like the "ephrine" that forms the ending of **phenylephrine** so you can relate the two decongestants.

## II. Allergic Rhinitis Steroid, Antitussives and Mucolytics

Allergic rhinitis is an inflammation (-itis) of the nose (rhin-). We treat it with a local steroid like **triamcinolone (Nasacort Allergy 24HR).** Nasal steroids don't work right away like a topical decongestant such as **oxymetazoline** (**Afrin**) might; rather, it takes weeks of use until a patient feels relief.

Because the name "**Robitussin**" has been associated with cough relief so long, many students mistakenly confuse plain **Robitussin** (just **guaifenesin**) with **Robitussin DM (guaifenesin / dextromethorphan)**. In **Robitussin DM**, the **guaifenesin** acts as a mucolytic to lyse (break up) mucous and chest congestion, while the **DM (dextromethorphan)** acts as the antitussive or cough suppressant. In severe cases, a **codeine**-based prescription product such as **guaifenesin / codeine (Cheratussin AC)** may be used.

## OTC Allergic Rhinitis Steroid Nasal Spray

**Triamcinolone (Nasacort Allergy 24HR)**
*try-am-SIN-oh-lone (NAY-zuh-cort)*

There is no recognized stem in **triamcinolone**, but often "o-n-e," pronounced like I "own" something, matches the "o-n-e" at the end of **testosterone**, a more familiar steroid. The brand name **Nasacort** reads like a story: "Nasa" for nose, "cort" for corticosteroid, "Allergy" for allergic rhinitis, and "24HR" for how long it works.

## OTC Antitussive / Mucolytic

**Guaifenesin / Dextromethorphan (Mucinex DM / Robitussin DM)**
*gwhy-FEN-uh-sin / decks-trow-meth-OR-fan (MEW-sin-ex dee-em, row-beh-TUSS-in dee-em)*

**Guaifenesin**, pronounced as if you put a "g" in front of "why," is a mucolytic, something that lyses or breaks up mucous. Students remember this because Mr. Mucus from the **Mucinex** commercials is green, which also starts with a "g." **Robitussin** "robs" your cough and "tussin" resembles "tussive." Antitussives are anti-cough medicines.

## RX Antitussive / Mucolytic

**Guaifenesin / Codeine (Cheratussin AC)**
*gwhy-FEN-uh-sin / CO-dean (CHAIR-uh-tuss-in ay-see)*

Sometimes **dextromethorphan (DM)** isn't enough and the patient needs **codeine** to suppress a cough.

Most students know **codeine**, but cough and **codeine** both start with "co" and that seems to help. The "chera" in **Cheratussin** comes from the product's cherry flavoring. Some are not sure what the "AC" means; some students think "anti-cough" al-though it's probably "and codeine."

## III. Asthma

**Asthma** is a disease of bronchoconstriction (the lung's branches tighten) and inflammation. For immediate relief during an attack, an **albuterol (ProAir HFA)** inhaler is the short-acting bronchodilator that will reverse the bronchoconstriction. **Proventil** is also a brand name for nebulized (a form of mist) **albuterol**. Oral steroids such as **methylprednisolone (Medrol)** and **prednisone (Deltasone)** help reduce lung inflammation after a severe attack.

The combination inhalers **fluticasone / salmeterol** (**Advair**) or **budesonide / formoterol (Symbicort)** provide relief from both inflammation and bronchoconstriction by combining a steroid and long-acting $beta_2$ receptor agonist (which bronchodilates). Notice that many steroids have "sone" not an official stem, at the end of their names, and that $beta_2$ receptor agonists have the stem "–terol" in the suffix. These long-acting combinations prophylactically prevent asthma attacks.

**Ipratropium** in **DuoNeb** and **Tiotropium (Spiriva)** are anticholinergic medications. Often these medications cause dry mouth, constipation and other unwanted adverse effects. However, **ipratropium** and **tiotropium (Spiriva)** affect the smooth muscle of the lungs, allowing for bronchodilation and relaxation of the bronchi. Both **albuterol** and **ipratropium** bronchodilate. One is an agonist of $beta_2$ receptors and the other is an antagonist of acetylcholine (ACh).

**Montelukast (Singulair)** inhibits leukotriene receptors. Leukotrienes cause bronchoconstriction, a process that protects the lungs against foreign contaminants. Drugs in this class end in "-lukast."

**Omalizumab (Xolair)** is an IgE antagonist and a biologic.

## ORAL STEROIDS

**Methylprednisolone (Medrol)**
*meth-ill-pred-NISS-uh-lone (MED-rol)*

A student connected "pred" and "predator of inflammation" for generic **methylprednisolone**. **Methylprednisolone** (**Medrol**) comes in a 6-day, 21-pill dose pack that gives patients 6 tablets on the first day, 5 on the 2nd, 4 on the 3rd, 3 on the 4th, 2 on the 5th, and 1 on the 6th. This dosing reminds us to taper steroids to allow the adrenal glands time to resume normal function.

**Prednisone (Deltasone)**
*PRED-ni-sewn (DEALT-uh-sewn)*

Most students know **prednisone**, but many steroid compounds have this unofficial "sone" ending. In this case, the "pred-" is the official stem and is in the prefix position. Note, we find this stem in the middle of the drug name in the steroid methyl**pred**nisolone.

## INHALED STEROID/BETA$_2$-RECEPTOR AGONIST LONG-ACTING

**Budesonide / Formoterol (Symbicort)**
*byou-DES-uh-nide / four-MOE-ter-all (SIM-buh-court)*

Similar to **fluticasone / salmeterol, budesonide** has the "sone" syllable in the middle of its name, and is pronounced "sone" even though it's spelled "s-o-n." The "-terol" in **formoterol** indicates a beta$_2$ agonist bronchodilator. You can think of the "S-y-m" in the brand name **Symbicort** as symbiotic, meaning "working with," plus "c-o-r-t" for corticosteroid.

**Salmeterol / Fluticasone (Advair)**
*flue-TIC-uh-sewn / Sal-ME-ter-all (ADD-vair)*

Recognize the steroid **fluticasone** by the unofficial "sone," and the long-acting bronchodilator **salmeterol** by the "-terol" stem. The brand name **Advair** seems like "add two drugs to get air."

## INHALED STEROID

**Fluticasone (Flovent HFA, Flovent Diskus, Flonase)**
*flue-TIC-uh-sewn (FLOW-vent)*

The "sone" ending, while not an official stem, is a useful clue that **fluticasone** is a steroid. In the past, inhalers used to feel cold when patients used them because the chlorofluorocarbon (CFC) propellant spray was similar to Freon, the refrigerant used in air conditioning. However, CFCs damage the ozone layer so the new hydrofluoroalkane (HFA) replaces the CFC propellant. The brand name **Flovent HFA** cleverly uses the first two letters of the generic name **fluticasone**, incorporates f-l-o from "airflow" and adds, "vent" to let the patient know this is administered in the mouth. The d-i-s-k-u-s inhaler is a device that looks like a d-i-s-c-u-s (the Frisbee-like thing used in the Olympics) that uses a dry powder for inhalation rather than a propellant and liquid. **Flonase** is the brand name for **fluticasone** administered nasally, and is available OTC.

## BETA2 RECEPTOR AGONIST SHORT-ACTING

**Albuterol (ProAir HFA)**
*Al-BYOU-ter-all (PRO-air aitch-ef-ay)*

The "-terol" stem of **albuterol** indicates it's a beta-2 adrenergic agonist that causes bronchodilation. Note, the "–terol" stem of **albuterol** does not help you know if it's long acting or short-acting; that distinction has to be memorized. The brand name **ProAir HFA** is straightforward with "Pro" as in "I'm for it" and "Air" for airway.

## BETA2 RECEPTOR AGONIST / ANTICHOLINERGIC SHORT-ACTING

**Albuterol / Ipratropium (DuoNeb)**
*Al-BYOU-ter-al / Ih-pra-TROPE-e-um (DUE-oh-neb)*

A "-terol" short-acting bronchodilator like **albuterol** can combine with a shorter acting anticholinergic like **ipratropium** (as compared to longer-acting **tiotropium**) in a nebulized form to treat asthma symptoms faster. Instructors use **atropine** as the prototype drug for the anticholinergics and you can see the "-trop-" stem in **atropine, ipratropium**, and **tiotropium**. The brand name **DuoNeb** indicates a duo of drugs in nebulized form.

## ASTHMA / COPD – ANTICHOLINERGIC LONG-ACTING

**Tiotropium (Spiriva)**
*tie-oh-TROW-pee-um (Spur-EE-va)*

**Tiotropium** has the same "-trop-" stem as **ipratropium** and is the long-acting version. As with the "-terols," the $beta_2$ receptor agonists, you have to memorize which anticholinergic is long-acting vs. short-acting. **Spiriva** takes the "spir" from respire.

## ASTHMA – LEUKOTRIENE RECEPTOR ANTAGONIST

**Montelukast (Singulair)**
*Mon-tee-LUKE-ast (SING-you-lair)*

Leukotrienes form in leukocytes (white blood cells) and cause inflammation. By blocking them, **montelukast** helps with the inflammatory component of asthma. The "-lukast" stem is very similar to leukotriene. The **Singulair** brand name comes from its once daily "single" dosing and the "air" it helps to bring into the asthmatic lungs.

## ASTHMA – ANTI-IGE ANTIBODY

**Omalizumab (Xolair)**
*oh-mah-liz-YOU-mab (ZOHL-air)*

Like **infliximab** for ulcerative colitis, **omalizumab** is a biologic. The "inf-" is a prefix that separates it from other similar drugs. The "-li-" stands for immunomodulator (the target), the "-zu-" stands for humanized (the source) and the "–mab" is for monoclonal antibody. There is a black box warning (a severe warning immediately at the beginning of the package insert) for the possibility of anaphylaxis after the first dose and even a year after the onset of treatment. Therefore, health providers inject **omalizumab** where a medicine for treating anaphylaxis is available. The brand name **Xolair** seems like "exhale" and "air."

## IV. Anaphylaxis

Anaphylaxis is a special type of allergic overreaction of the body to something like an insect bite or bee sting. **Epinephrine (EpiPen)** quickly reverses the reaction, keeping the airway open.

**Epinephrine (EpiPen)**
*eh-peh-NEF-rin (EP-ee-pen)*

The word **epinephrine** has a Greek origin. "Epi" means "above," and "neph" means "kidney." Above the kidney is the adrenal gland responsible for the body's natural release of **epinephrine**. The Latinized version of **epinephrine** is adrenaline. The "ad" means "above" and "renal" means "kidney." The **EpiPen** brand name comes from the injector device that looks somewhat like a pen.

## RESPIRATORY DRUG QUIZ (LEVEL 1)

Classify these drugs by placing the corresponding drug class letter next to each medication. Try to underline the stems before you start and think about the brand name and function of each drug.

1. Albuterol (ProAir)
2. Cetirizine (Zyrtec)
3. Diphenhydramine (Benadryl)
4. Fluticasone/salmeterol (Advair)
5. Guaifenesin/DM (Robitussin DM)
6. Tiotropium (Spiriva)
7. Loratadine (Claritin)
8. Montelukast (Singulair)
9. Pseudoephedrine (Sudafed)
10. Methylprednisolone (Medrol)

**Respiratory drug classes:**

A. 1st-generation antihistamine
B. 2nd-generation antihistamine
C. Anticholinergic
D. Decongestant
E. Leukotriene receptor antagonist
F. Mucolytic/cough suppressant
G. Oral steroid
H. Nasal steroid
I. Short-acting bronchodilator
J. Steroid / long-acting bronchodilator

## Respiratory drug quiz (Level 2)

Classify these drugs by placing the corresponding drug class letter next to each medication. Try to underline the stems before you start and remember the brand name and function of each drug.

1. Budesonide / formoterol
2. Guaifenesin / codeine
3. Triamcinolone
4. Fluticasone / salmeterol
5. Pseudoephedrine
6. Cetirizine
7. Diphenhydramine
8. Albuterol
9. Montelukast
10. Prednisone

**Respiratory drug classes:**

A. 1st-generation antihistamine
B. 2nd-generation antihistamine
C. Anticholinergic
D. Decongestant
E. Leukotriene receptor antagonist
F. Mucolytic / cough suppressant
G. Oral steroid
H. Nasal steroid
I. Short-acting bronchodilator
J. Steroid / long-acting bronchodilator

## RESPIRATORY: MEMORIZING THE CHAPTER

| | | |
|---|---|---|
| **Diphenhydramine** | **Triamcinolone** | Fluticasone |
| **Cetirizine** | **Guaifenesin/DM** | Albuterol |
| **Loratadine** | Guaifenesin/codeine | Albuterol/Ipratropium |
| **Loratadine-D** | Methylprednisolone | Tiotropium |
| **Pseudoephedrine** | Prednisone | Montelukast |
| **Phenylephrine** | Budesonide/Formoterol | Omalizumab |
| **Oxymetazoline** | Fluticasone/Salmeterol | Epinephrine |

### "R" RESPIRATORY

Start with the 1st-generation OTC antihistamine **diphenhydramine**, and then go to the 2nd-generation **cetirizine** (C) alphabetically to 2nd-generation **loratadine** (L). Add a decongestant to make **loratadine-D** behind-the-counter (BTC), and then move to what the "hyphen D" stands for – **pseudoephedrine** the decongestant. Then walk into the OTC aisle to get the oral decongestant **phenylephrine** and move up to the nose with the intranasal decongestant **oxymetazoline**. Stay in the nose with an OTC intranasal glucocorticoid **triamcinolone**. Move from nasal congestion to chest congestion with **guaifenesin / dextromethorphan** to another antitussive combination behind-the-counter, **guaifenesin with codeine**. If that does not work and the coughing inflames your lungs, you might need an oral steroid like **methylprednisolone** (M) or **prednisone** (P). After the acute attack, you might find you have asthma and need to use a prophylactic inhaled steroid in combination with a long-acting beta$_2$ receptor agonist, either **budesonide / formoterol** or **salmeterol / fluticasone**. You could individually use the steroid

**fluticasone** or **albuterol**, the short-acting beta$_2$ receptor agonist. The beta$_2$ receptor agonist **albuterol** combined with anticholinergic **ipratropium** makes the duo in **DuoNeb**, or alternatively, the long-acting anticholinergic **tiotropium** can be given alone. If that doesn't work, use **montelukast** against leukotrienes or **omalizumab** against IgE, but remember that **omalizumab** has a black box warning about anaphylaxis, which would necessitate an injection of **epinephrine**.

# CHAPTER 4 IMMUNE

## I. OTC Antimicrobials

Antimicrobials, meaning "against microbes," can generally be divided into three classes: **antibiotics** (drugs for bacteria), **antifungals** (drugs for mycoses or fungi), and **antivirals** (drugs for viruses). Antibiotic brand names don't give good information about drugs because they derive from generic names. As such, creating linkages between drug classes so you can group similar antibiotics becomes critical. For example, penicillins and cephalosporins, along with **vancomycin**, affect bacterial cell walls. By putting them near each other on this list, you can group them into a larger category based on function.

Antifungals and antivirals, in contrast, have brand names that allude to their therapeutic effect. Most students try to be ultra-efficient and just memorize generic names. Just as you have more information if you know a first and last name, you know more about a drug by memorizing both generic and brand names. This backup information is critical under the stress of exams or clinical practice.

### Antibiotic cream

**Neomycin / Polymyxin B / Bacitracin (Neosporin)**
*knee-oh-MY-sin / pall-EE-mix-en / bah-seh-TRACE-in (KNEE-oh-spore-in)*

> **Neomycin** is an aminoglycoside that is generally toxic to the kidney (nephrotoxic) and ears (ototoxic)

when used systemically. However, patients can safely use topical preparations containing **neomycin** such as over-the-counter **Neosporin**. The brand **Neosporin** takes "N-e-o" from **neomycin**, "p-o" from **polymyxin B**, and "r-i-n" from **bacitracin.**" We associate "spores" with fungi, and this helps link **Neosporin** to fungal infections.

## ANTIFUNGAL CREAM

**Butenafine (Lotrimin Ultra)**
*BYOO-ten-uh-feen (LOW-treh-min)*

**Butenafine** treats topical fungal infections like ringworm, jock itch, and athlete's foot. Sometimes you will see the Latin names for these conditions: tinea coporis (ringworm), tinea cruris (jock itch), and tinea pedis (athlete's foot) respectively.

## ANTIVIRAL (PROPHYLAXIS)

**Influenza Vaccine (Fluzone, Flumist)**
*in-FLU-en-zah VACK-seen (FLEW-zone, FLEW-mist)*

While some children might need a prescription for the **influenza vaccine**, adults can walk up to the pharmacy counter and get a flu shot at certain pharmacies. Be careful; generic names with "f-l-u," for example, **fluconazole,** an antifungal, and **fluoxetine**, an antidepressant, refer to a fluorine atom in their chemical structure, not to the influenza virus. The "flu" in the vaccine brand names **Fluzone** and **Flumist** indicates influenza, just like **Tamiflu**, an oral medication for influenza infection contains the syllable "flu." The nasal vaccine **Flumist** provide an alternative for patients who don't want an injection.

### ANTIVIRAL (ACUTE)

**Docosanol (Abreva)**
*Do-cah-SAN-all (uh-BREE-vah)*

**Docosanol** is a topical antiviral for cold sores caught and treated early. I thought who would pay twenty dollars for a small tube like that? Then I thought of homecoming dances. So, use **docosanol**, so you can go to the ball. **Abreva**, the brand name, hints at therapeutic effect as **docosanol** "abbreviates" the time a cold sore lasts.

## II. ANTIBIOTICS AFFECTING CELL WALLS

Bacteria have cell walls. Human cells don't (although they do have cell membranes). This introduces *selectivity*. If a drug targets a tissue or structure that bacteria have but humans don't, it should be selective for the bacteria and safe for the patient.

**Penicillins** were one of the first antimicrobial classes discovered. **Penicillin's** mechanism of action is to open a bacterium's cell wall, like popping a bubble, to kill it. This killing action is termed bactericidal. However, sometimes we see resistance to a single antibiotic like **amoxicillin (Amoxil).** For example, a child with an ear infection finishes a course of "the pink stuff" and remains sick. **Amoxicillin / clavulanate (Augmentin)** adds the compound clavulanate to protect the **amoxicillin** against an enzyme bacteria produce called a beta-lactamase. The enzyme acquired its name from the chemical structure (a beta lactam) that's in all penicillins. This additional component, **clavulanate**, helps **amoxicillin** work in cases where it had previously failed.

**Cephalosporins** can have cross-sensitivity with penicillins. Patients allergic to one may be allergic to the other, but this is quite rare. We classify cephalosporins into generations. The first-generation drugs, such as **cephalexin (Keflex)**, don't penetrate the cerebrospinal fluid (CSF), have poor gram-negative bacterial coverage (gram-negative bacteria have an extra protective layer and do not take up a gram stain), and are subject to deactivation by beta-lactamase producing bacteria. As we move to third-generation **ceftriaxone (Rocephin)** and fourth-generation **cefepime (Maxipime)**, we get good penetration into the CSF, good gram-negative coverage and the antibiotics cover bacteria resistant to beta-lactam drugs.

**Vancomycin (Vancocin)** is sometimes the last line of defense against a sometimes-deadly bacterial infection like methicillin-resistant *Staphylococcus aureus* (MRSA). To minimize resistance, a special protocol dictates who can and cannot get **vancomycin.** In rare cases, **vancomycin** can cause a hypersensitivity reaction called red man syndrome. **Vancomycin** has special dosing requirements for patient safety and often pharmacists use their expertise to dose it appropriately so that patients receive optimal drug therapy.

## ANTIBIOTICS: PENICILLINS

### Amoxicillin (Amoxil)

*uh-mocks-eh-SILL-in (uh-MOCKS-ill)*

**Amoxicillin** has the "–cillin" stem that indicates its relationship to the penicillin family. The "a-m-o" probably came from the fact that it's an "amino" penicillin. The "-cillin" stem sounds like "cell-in" and can help you remember that **amoxicillin** or, more generally, **penicillins** destroy the cell wall. The brand name **Amoxil** simply removes an "i-c" and "l-i-n" from the generic name.

## Penicillin / Beta-lactamase inhibitor

**Amoxicillin / Clavulanate (Augmentin)**
*uh-mocks-eh-SILL-in / clav-you-LAN-ate (awg-MENT-in)*

When **amoxicillin** alone doesn't work, **Augmentin** augments **amoxicillin's** defenses against the bacterial beta-lactamase enzyme with **clavulanate**. I think of the "clavicle," the bone in your shoulder, as protective of the upper lung and associate **clavulanate** with that same protective effect.

## Cephalosporins

**Cephalexin (Keflex)**
*sef-uh-LEX-in (KE-flecks)*

With cephalosporins, a newer generation has better properties than the last, relative to what the prescriber is treating. Those advantages include better penetration into the cerebrospinal fluid (CSF), better gram- negative coverage, and better resistance to beta-lactamases. You may lose some gram-positive coverage as you move up the spectrum, however. The "ceph-" is an old stem from the first generation. The new stem "cef-" identifies the newer generations. That's how I remember **cephalexin** as being first generation. The brand name **Keflex** takes some letters from **cephalexin** to make its name.

**Ceftriaxone (Rocephin)**
*sef-try-AX-own (row-SEF-in)*

In the generic name, **ceftriaxone's** "cef-" indicates it's a cephalosporin. There is a "tri" in the generic name that you can use to remember it's 3rd generation. **Rocephin**, the brand name, seems to come from Hoffman-LaRoche's patent. The drug company took the "Ro" from "LaRoche," and the "ceph" and "in" from cephalosporin to make **Ro-ceph-in**.

**Cefepime (Maxipime)**
*SEF-eh-peem (MAX-eh-peem)*

**Cefepime** is a fourth-generation cephalosporin. I've remembered it by thinking of four letters "p-i-m-e" that are in both the brand name **Maxipime** and the generic name **cefepime**. Also, at the time, **Maxipime** was the **maximum** generation, the fourth and highest. However now there is a fifth-generation cephalosporin.

## GLYCOPEPTIDE

**Vancomycin (Vancocin)**
*van-co-MY-sin (VAN-co-sin)*

**Vancomycin's** "–mycin" stem isn't very useful for finding its therapeutic class. All it really means is that chemists derived **vancomycin** from the *Streptomyces* bacteria. I remember the function as "**vancomycin** will vanquish MRSA." To remember the brand name **Vancocin**, remove the "my" from **vancomycin**.

## III. Antibiotics – Protein synthesis inhibitors – bacteriostatic

We name bacteriostatic **tetracyclines** like **doxycycline (Doryx)** and **minocycline (Minocin)** after the four (tetra) member chemical ring (cycline). Tetracyclines and fluoroquinolones both cause photosensitivity and chelation (binding with cations such as the Ca++ in milk or antacids).

We sometimes call **macrolides** "**erythromycins**" after one of the original drugs in the class. Patients take **azithromycin** (**Zithromax**) as a double dose on the first day and a single dose the following four days. The double dose is a *loading dose*. Once-daily dosing improves patient compliance. Patients take **clarithromycin (Biaxin)** twice a day - notice the "bi" prefix in the brand name, and we dose **erythromycin (E-Mycin**) four times a day.

Dentists use **clindamycin (Cleocin)** for dental prophylaxis when a patient is penicillin allergic. Patients use it topically for severe acne. When used orally, it can cause a severe condition known as pseudomembranous colitis, also known as antibiotic-associated diarrhea (AAD).

**Linezolid (Zyvox)** is an oxazolidinone antibiotic that can work on both **methicillin**-resistant *Staphylococcus aureus* (MRSA) and **vancomycin**-resistant enterococci (VRE).

### Tetracyclines

**Doxycycline (Doryx)**
*docks-ee-SIGH-clean (DOOR-icks).*

I use the "d" in **doxycycline** to remind me that dentists use it to treat periodontal disease. **Doryx,** the brand name, takes the first four letters of **doxycycline** and adds an "r."

**Minocycline (Minocin)**
*MIN-oh-SIGH-clean (MIN-oh-sin)*

**Minocycline**, like **doxycycline** has the **tetracycline** class "-cycline" stem. To create the brand name **Minocin**, the manufacturer dropped the "c-y-c-l" and last "e."

## Macrolides

**Azithromycin (Zithromax Z-Pak)**
*ay-zith-row-MY-sin" (ZITH-row-max)*

To recognize the three macrolides, **azithromycin**, **clarithromycin**, and **erythromycin**, you will see the "–mycin" ending, but also a possible infix of "thro-." Be careful: there are macrolides without this infix and stem. The brand name **Zithromax** takes seven letters from **azithromycin** to construct its name. The **Zithromax Z-pak** is a convenient six-tablet package that includes a five-day course, two tablets for a loading dose on day one and one tablet for each thereafter.

**Clarithromycin (Biaxin)**
*Claire-ITH-row-my-sin (bi-AX-in)*

Gastroenterologists prescribe **clarithromycin** for peptic ulcer disease (PUD) triple therapy along with **amoxicillin** and a proton pump inhibitor like **omeprazole**. The **Biaxin** brand name indicates the twice daily dosing from the Latin abbreviation b.i.d. or *bis in die*.

**Erythromycin (E-Mycin)**
*err-ith-row-MY-sin (E-MY-sin)*

Some **erythromycin** tablets are bright red and that might be where it got its name. An erythrocyte is a red blood cell, and the word comes from connecting "erythro," the Greek for "red," and "cyte," for cell. The brand name **E-mycin** comes from taking the "rythro" out of the generic name **erythromycin**.

## LINCOSAMIDE

**Clindamycin (Cleocin)**
*clin-duh-MY-sin (KLEE-oh-sin)*

Most students remember the adverse effect CDAD (*Clostridium difficile*-Associated Disease) because there is a "c" and a "da" right after in the generic **clindamycin**. To make the brand name **Cleocin,** the manufacturer replaced the "i-n-d-a-m-y" in **clindamycin** with "e-o."

## OXAZOLIDINONE

**Linezolid (Zyvox)**
*LYNN-ez-oh-lid (ZIE-vocks)*

The "-zolid" stem in **linezolid** comes from the **oxazolidinone** class. I think it's more helpful to think, "Man, **Zyvox** is zolid (solid); it treats two very difficult to treat organisms, MRSA and VRE.

# IV. Antibiotics – Protein synthesis inhibitors – Bactericidal

Bactericidal **aminoglycosides** can damage the kidneys (nephrotoxicity) and ears (ototoxicity). I think of the "side" in **aminoglycoside** and "cide" as in "cidal" to remind me these are killers.

## Aminoglycosides

**Amikacin (Amikin)**
*am-eh-KAY-sin (AM-eh-kin)*

> Some internet sources say that a "cin" ending means an aminoglycoside, but that's not necessarily true. Many antibiotics end in "c-i-n," so I think it's more useful to look at the "a-m-i" that is in the words **aminoglycoside**, **amikacin** and **Amikin**. The brand name **Amikin** is simply **amikacin** without the "a-c."

**Gentamicin (Garamycin)**
*Jenn-ta-MY-sin (gare-uh-MY-sin)*

> Just as practitioners abbreviate **vancomycin** as "vanc" in conversation, they abbreviate **gentamicin** as "gent." The brand name **Garamycin** is similar to **gentamicin** spelled with "**-mycin**" not "**-micin.**"

# V. Antibiotics for urinary tract infections (UTIs) and peptic ulcer disease (PUD)

**Sulfamethoxazole / trimethoprim (Bactrim)** is a combination therapy that affects the folic acid in bacteria. Humans can safely ingest folic acid, so it doesn't affect us adversely. However, sulfa medications can sometimes cause allergic reactions. **Sulfamethoxazole** can even cause a rare but life-threatening condition of the skin and mucous membranes known as Stevens-Johnson syndrome. Sulfa drugs clear urinary tract infections (UTIs) and provide prophylaxis (prevention) of certain infections that commonly occur in immunocompromised patients such as HIV patients.

We sometimes call **fluoroquinolones** "floxacins" after their infix "-fl-" + suffix "-oxacin." Like tetracyclines, fluoroquinolones cause photosensitivity (an increased sensitivity to burning from sunlight) and chelation (a binding with cations such as the Ca++ in milk or antacids). **Fluoroquinolones** have a very unusual side effect in that sometimes they can cause tendon rupture, although rarely.

**Metronidazole (Flagyl)** treats various infections, including *H. Pylori,* as part of triple therapy. A notable side effect of **metronidazole** is the disulfiram reaction where a patient may experience serious nausea and vomiting. Projectile vomiting is rare, but a vivid way to remember **metronidazole's** adverse effect with alcohol.

## DIHYDROFOLATE REDUCTASE INHIBITORS

**Sulfamethoxazole / Trimethoprim (SMZ-TMP)**
*sull-fa-meth-OX-uh-zol e /try-METH-oh-prim*
*(ess-em-zee / tee-em-pee)*

> **SMZ / TMP** is the acronym for **sulfamethoxazole / trimethoprim**. I want to caution you about seeing sulfa in the name and allergic reactions. While sulfa drugs have "s-u-l-f-a" in them, some drugs have sulfa groups in the chemicals structure, but not in

the generic name, e.g., **furosemide.** The academic literature doesn't support cross-sensitivity between allergies to sulfa antibiotics and other sulfonamide containing drugs like **furosemide**. The brand name **Bactrim** contains "b-a-c-t-r-i-m" from "**bact**e**r**i**u**m."

## FLUOROQUINOLONES

**Ciprofloxacin (Cipro)**
*sip-row-FLOCKS-uh-sin (SIP-row)*

The **quinolone** stem is "-oxacin," but **fluoroquinolones** have the "-fl-" infix also. One student remembered quinolones are for UTIs because Dr. Quinn, Medicine Woman, is female. Women get proportionally more UTIs than men do. By cutting the "–floxacin" stem from the generic name **ciprofloxacin**, the manufacturer made the brand name, **Cipro**.

**Levofloxacin (Levaquin)**
*Lee-vo-FLOCKS-uh-sin (LEV-uh-Quinn)*

**Levofloxacin** is the left-handed (levo) isomer of **ofloxacin**, another **fluoroquinolone**. The brand name combines the "lev" from levofloxacin and "quin" from fluoroquinolone to form **Levaquin**.

## NITROIMIDAZOLE

**Metronidazole (Flagyl)**
*met-ruh-NYE-duh-zole (FLADGE-ill)*

The generic name **metronidazole** contains one of the three "i's" from nitroimidazole, its parent class.

Gastroenterologists use **metronidazole** for peptic ulcer disease (PUD). Note that **metronidazole** is technically an antiprotozoal and students look at the "azole" ending, which is a little similar to "ozoal" from "protozoal." A student learned to give "Flag,"a shorter form of **Flagyl**, for *B. frag* a shortening of the *Bacteroides fragilis* infections.

## VI. ANTI-TUBERCULOSIS AGENTS

Prescribers use anti-tuberculosis agents for an extended period (several months) because tuberculosis organisms grow slowly. Multiple drug therapy helps prevent resistance. I use the acronym "r-i-p-e," to remember the four major antituberculosis agents: **rifampin**, **isoniazid**, **pyrazinamide**, and **ethambutol**. Non-drug resistant, non-HIV patients take all four drugs for two months, and then **isoniazid** and **rifampin** together for four more months.

**Rifampin (Rifadin)**
*rif-AM-pin (rif-UH-din)*

Students remember that **rifampin** turns secretions like tears, sweat, and urine red with its first letter "r." The brand name **Rifadin** simply replaces the "m-p" from **rifampin** with a "d."

**Isoniazid (INH)**
*eye-sew-NIGH-uh-zid (EYE-en-aitch)*

There is no "H" in **isoniazid**, so **INH** comes from the chemical name isonicotinylhydrazide. The "n-i" in **isoniazid** reminds students that peripheral neuritis is an adverse effect.

**Pyrazinamide (PZA)**
*pier-uh-ZIN-uh-mide (pee-zee-ay)*

> The "p" in **pyrazinamide** reminds students that an adverse effect is polyarthritis. **Pyrazinamide's** abbreviation is **PZA**.

**Ethambutol (Myambutol)**
*eh-THAM-byou-tall (my-AM-byou-tall)*

> The "e" for "eyes" or "o" in **ethambutol** helps remind students that optic neuritis is an adverse effect. To make the brand name, the manufacturer replaced the "eth" in **ethambutol** with "my" in **Myambutol** because *Mycobacterium tuberculosis* is the causative agent.

## VII. Antifungals

Scientists divide antifungals into two general types: systemic (in the body) and dermatologic or topical (on the skin). Before the advent of antifungals, most systemic fungal infections were deadly. **Amphotericin B (Fungizone)** can treat systemic infections. **Fluconazole (Diflucan)** orally treats vaginal yeast infections. **Nystatin (Mycostatin)** can eliminate thrush or yeast infections.

**Amphotericin B (Fungizone)**
*am-foe-TER-uh-sin bee (FUN-gah-zone)*

> What about **amphotericin A**? Well, it didn't do anything, so they came up with **amphotericin B**. While the antibacterials' brand names didn't do a very good job helping to indicate their therapeutic effects, this antifungal's brand name, **Fungizone**, makes it easier to know its therapeutic use.

**Fluconazole (Diflucan)**
*flue-CON-uh-zole (die-FLUE-can)*

The "–conazole" ending helps identify **fluconazole** as an antifungal drug. Again, be careful of the "f-l-u" in **fluconazole**, which is for a fluorine atom it contains, not influenza. One student came up with using the first three letters of the brand name **Diflucan** as "Die fungi!"

**Nystatin (Mycostatin)**
*NIGH-stat-in (MY-co-stat-in)*

**Nystatin** is an interesting generic name because it ends in "statin." A class of cholesterol lowering drugs, the HMG-CoA reductase inhibitors, commonly referred to as "statins," have a similar ending. A better infix + suffix stem for HMG-CoA reductase inhibitors is "vastatin." To keep from thinking **nystatin** was ever a cholesterol lowering "statin," one student remembered the dosage forms nystatin comes in: powder and liquid to swish / spit / swallow. **Mycostatin**, the brand name, dropped the "ny" from **nystatin** and added "Myco," a prefix often seen with mycoses (fungal infections).

## VIII. Antivirals – Non-HIV

Many antivirals have "-vir-" in the middle or at the end of the generic and / or brand name. Drugs for influenza, such as **oseltamivir (Tamiflu)** and **zanamivir (Relenza)** work when taken within 48 hours of the infection. Drugs for herpes infections such as **acyclovir (Zovirax)** and **valacyclovir (Valtrex)** can help prevent recurrences and treat an infection, but they do not cure the disease.

Respiratory syncytial virus (RSV) is usually unproblematic in healthy adults, but in infants younger than one year old, it can be deadly. A drug like the vaccine **palivizumab (Synagis)** can prevent RSV in at-risk patient populations.

## INFLUENZA A AND B

**Oseltamivir (Tamiflu)**
*owe-sell-TAM-eh-veer (TA-mi-flue)*

> Often family members will all get prescriptions for **oseltamivir** if one person is sick enough or if a family member is immunocompromised. The brand name **Tamiflu** alludes to a drug that "tames the flu." It's prescribed for acute influenza or prophylaxis.

**Zanamivir (Relenza)**
*za-NAH-mi-veer (rah-LEN-zuh)*

> **Zanamivir** comes in a Diskhaler, a way to get powder to the lungs. The Diskhaler is difficult for patients with dexterity issues, but provides an alternative to **oseltamivir (Tamiflu).** Think: **Relenza** "**re**presses influ**enza**" virus or **Relenza** makes "influ**enza rel**ent" (give up).

## HERPES SIMPLEX VIRUS & VARICELLA-ZOSTER VIRUS (HSV/VSV)

**Acyclovir (Zovirax)**
*ay-SIGH-clo-veer (zo-VIE-racks)*

> **Zovirax** treats Varicella-Zoster virus (VSV) and herpes simplex virus (HSV). You can think of

**Zovirax** as a drug that axes Zoster virus. Dosing is five times daily.

**Valacyclovir (Valtrex)**
*Val-uh-SIGH-clo-veer (VAL-trex)*

**Valacyclovir** has **acyclovir** in the name because it's the valine ester. A prodrug like **valacyclovir** turns into an active drug in the body, in this case, **acyclovir. Valacyclovir** allows for twice daily dosing, so prescribers prefer the oral form of **valacyclovir** to **acyclovir** for patient compliance. The brand name, **Valtrex,** includes the "val" from **valacyclovir** plus T-rex, and wrecks a virus.

### Respiratory syncytial virus (RSV)

**Palivizumab (Synagis)**
*pal-eh-viz-YOU-mab (SIN-uh-giss)*

The prefix "p-a-l-i" has "P" and "I" in it. You can remember **palivizumab** is for pediatrics or infants at risk for RSV. In **palivizumab**, the "pali" is a prefix that separates it from similar drugs. The "-vi-" stands for antiviral (the target), the "-zu-" stands for humanized (the source), and the "–mab" is for monoclonal antibody. This biologic stem + infixes resemble **infliximab (Remicade)** for ulcerative colitis or **omalizumab (Xolair)** for asthma, but with a different clinical purpose.

## VIII. Antivirals – HIV

HIV drugs affect specific targets in the cell or retrovirus. HIV medications, like tuberculosis medications, often work best in drug combinations. I've organized the five HIV drug classes in the order an HIV virus attacks a healthy cell. First, the HIV virus tries to fuse with the cell, then it uses cellular chemokine receptor five (CCR5) to enter the cell. Inside the cell, the HIV virus uses reverse transcriptase, integrase, and protease. HIV medications have three letter abbreviations, as these drugs are not only hard to pronounce, but conversation filled with several multisyllable words can make comprehension difficult.

## FUSION INHIBITOR

**Enfuvirtide (Fuzeon) (T-20)**
*En-FYOO-veer-tide (FYOO-zee-on)*

It's easier to remember **enfuvirtide's** brand name **Fuzeon** first because it's a fusion inhibitor. Inside the generic name, you see "vir" for antiviral and we pronounce the "f-u" as FYOO. Put that together and you can remember **enfuvirtide** is **Fuzeon**, a fusion inhibitor. I use the "T" in "**T-20**" to remember that "tide" is the last syllable in the generic name.

## CELLULAR CHEMOKINE RECEPTOR (CCR5) ANTAGONIST

**Maraviroc (Selzentry) (MVC)**
*MARE-uh-VIR-ock (SELLS-en-tree)*

The stem "-vir-" is inside the generic name **maraviroc**. The sub-stem "-viroc" has five letters, with the "c" at the end of the generic name, so you can remember it's a CCR5 antagonist. You can think of **maraviroc** as a "rock" guarding against viral entry. The brand name **Selzentry** sounds a lot like "sentry," someone who guards.

## NON-NUCLEOSIDE REVERSE TRANSCRIPTASE INHIBITORS (NNRTIS) WITH 2 NUCLEOSIDE / NUCLEOTIDE REVERSE TRANSCRIPTASE INHIBITORS (NRTIS)

**Efavirenz / Emtricitabine / Tenofovir (Atripla) (EFV / FTC / TDF)**
*eh-FAH-vir-enz / EM-try-SIGH-tah-been / ten-OFF-oh-vir (ay-TRIP-lah)*

When you see something complex to memorize, first look at the sub-class of antiviral. Instead of trying to memorize the whole name, try to memorize the stems "-virenz," "-citabine," and "-vir." Then add the other two or three syllables to memorize the whole names of **efavirenz, emtricitabine**, and **tenofovir**. The brand name **Atripla** can be thought of as three drugs, "triple" surrounded by two A's that can stand for "against AIDS."

## INTEGRASE STRAND TRANSFER INHIBITOR

**Raltegravir (Isentress) (RAL)**
*ral-TEG-ra-veer (EYE-sen-tress)*

**Raltegravir** is an integrase strand transfer inhibitor. Inside the generic name, you can find the stem "-tegravir" made up of "tegra," a part of integrase and "vir" for antiviral. The brand name **Isentress** also looks like sentry, except it has the "I" to remind you of integrase.

## PROTEASE INHIBITOR

**Darunavir (Prezista) (DRV)**
*dar-YOU-nah-veer (Pre-ZIST-uh)*

The brand name, in this case, is a little easier. **Prezista** sounds like resist spelled "r-e-z-i-s-t," and the first two letters "p-r" of "protease."

## IMMUNE DRUG QUIZ (LEVEL 1)

Classify these drugs by placing the corresponding drug class letter next to each medication. Try to underline the stems before you start and think about the brand name and function of each drug.

1. Amoxicillin (Amoxil)
2. Azithromycin (Zithromax)
3. Cefepime (Maxipime)
4. Ceftriaxone (Rocephin)
5. Fluconazole (Diflucan)
6. Gentamicin (Garamycin)
7. Isoniazid (INH)
8. Levofloxacin (Levaquin)
9. Nystatin (Mycostatin)
10. Valacyclovir (Valtrex)

**Immune drug classes:**

A. 1st-generation cephalosporin
B. 2nd-generation cephalosporin
C. 3rd-generation cephalosporin
D. 4th-generation cephalosporin
E. Antibiotic: aminoglycoside
F. Antibiotic: fluoroquinolone
G. Antibiotic: macrolide
H. Antibiotic: penicillin
I. Antibiotic: sulfa
J. Antibiotic: tetracycline
K. Anti-fungal
L. Anti-tuberculosis
M. Anti-viral (herpes)
N. Anti-viral (HIV)
O. Anti-viral (influenza)

## Immune drug quiz (Level 2)

Classify these drugs by placing the corresponding drug class letter next to each medication. Try to underline the stems before you start and remember the brand name and function of each drug.

1. Rifampin
2. Amphotericin B
3. Amikacin
4. Ciprofloxacin
5. Pyrazinamide
6. Acyclovir
7. Cephalexin
8. Sulfamethoxazole / Trimethoprim
9. Erythromycin
10. Oseltamivir

**Immune drug classes:**

A. $1^{st}$-generation cephalosporin
B. $2^{nd}$-generation cephalosporin
C. $3^{rd}$-generation cephalosporin
D. $4^{th}$-generation cephalosporin
E. Antibiotic: aminoglycoside
F. Antibiotic: fluoroquinolone
G. Antibiotic: macrolide
H. Antibiotic: penicillin
I. Antibiotic: sulfa
J. Antibiotic: tetracycline
K. Anti-fungal
L. Anti-tuberculosis
M. Anti-viral (herpes)
N. Anti-viral (HIV)
O. Anti-viral (influenza)

## IMMUNE: MEMORIZING THE CHAPTER

| **Neomycin /** | Azithromycin | Amphotericin B |
|---|---|---|
| **Polymyxin-B /** | Clarithromycin | Fluconazole |
| **Bacitracin** | Erythromycin | Nystatin |
| **Butenafine** | Clindamycin | Oseltamivir |
| **Influenza vaccine** | Linezolid | Zanamivir |
| **Docosanol** | Amikacin | Acyclovir |
| Amoxicillin | Gentamicin | Valacyclovir |
| Amoxicillin / | Sulfamethoxazole / | Palivizumab |
| Clavulanate | Trimethoprim | Enfuvirtide (T-20) |
| Cephalexin | Ciprofloxacin | Maraviroc (MVC) |
| Ceftriaxone | Levofloxacin | Efavirenz / |
| Cefepime | Metronidazole | Emtricitabine / |
| Vancomycin | Rifampin | Tenofovir |
| Doxycycline | Isoniazid (INH) | (EFV/FTC/TDF) |
| Minocycline | Pyrazinamide (PZA) | Darunavir (DRV) |
|  | Ethambutol | Raltegravir (RAL) |

### "I" IMMUNE

The sub-algorithm is to start with antibacterial agents, then go to antifungals, then antivirals, which are in alphabetical order b-f-v. So first the OTC antibacterial cream **neomycin / polymyxin-B / bacitracin**; **butenafine**, the antifungal cream; and the prophylactic antiviral **influenza vaccine**; followed by an acute antiviral – **docosanol**. Begin again with systemic antibacterials, first those that attack the cell wall, the beta lactamase susceptible amino penicillin **amoxicillin**, followed by the augmented beta lactamase resistant **amoxicillin / clavulanate.** Move to cephalosporins in generational order: 1st generation **cephalexin**; to 3rd generation **cef*tri*axone**; to 4th generation **cefepime**, the "maximum" generation; to alphabetically last "v" for glycopeptide **vancomycin** for MRSA. From bactericidal cell wall attackers to bacteriostatic inhibitors of protein synthesis, the tetracyclines: **doxycycline** and **minocycline** followed by three macrolides in order of the number of times taken per day: qd (once daily), bid (twice daily), qid (four times daily), **azithromycin**, **clarithromycin**, and **erythromycin** respectively. Use the "-mycin" ending to get to **clindamycin** and the "l-i-n" from clindamycin to go to **linezolid**. Follow with the bactericidal inhibitors of protein synthesis, the aminoglycosides **amikacin** and **gentamicin**. Move to the UTI medications **sulfamethoxazole** and **trimethoprim**, **ciprofloxacin**, and **levofloxacin**; and from "l" in "levo" to "m" in **metronidazole**, the antiprotozoal. Then four letters for four TB drugs because over 4 mm is a positive TB test or a ripe, "r-i-p-e" result: **rifampin** (R), **isoniazid** (I), **pyrazinamide** (P), and **ethambutol** (E). TB patients often have opportunistic fungi, so three antifungals alphabetically follow: **amphotericin B**, **fluconazole**, and **nystatin.** The antivirals for influenza follow those: **oseltamivir** and **zanamivir**; then **acyclovir** and **valacyclovir** for HSV, in order of half-life; then for RSV **palivizumab**; then for HIV, in order of attack: fusion, CCR5, reverse transcriptase, integrase, and protease. I have to use brand names first because those are easier: **Fuzeon, Selzentry, Atripla, Isentress, Prezista**, and then the generic names for those: **enfuvirtide**, **maraviroc**, (**efavirenz** / **emtricitabine** / **tenofovir**), **raltegravir** and **darunavir**.

# CHAPTER 5 NEURO

## I. OTC Local Anesthetics and Antivertigo

There are two major classes of local anesthetics named after the molecules in the middle of their structures: esters and amides. Esters, such as **benzocaine (Anbesol),** are generally found in topical agents because when given by injection, they are more allergenic (cause allergic reactions). Amides are less allergenic, therefore, **lidocaine**, an amide, is usually well tolerated when injected. Most students know of **cocaine** and remember these amides are local anesthetics through the "-caine" ending association among the three names: **benzocaine, lidocaine**, and cocaine. The over-the-counter (OTC) antiemetic / motion sickness medicine **meclizine** (**Dramamine**) also has a brand name **Antivert**, for <u>anti</u>-<u>vert</u>igo, which helps memorization.

### Local Anesthetics – Ester Type

**Benzo<u>caine</u> (Anbesol)**
*BEN-zoh-cane (ANN-buh-sawl)*

The "-caine" stem indicates **benzocaine** is a local anesthetic. **Anbesol** <u>n</u>um<u>bs</u> an aching tooth.

### Local Anesthetics – Amide Type

**Lido<u>caine</u> (Solarcaine)**
*LIE-doe-cane (SOH-ler-cane)*

**Lidocaine** is often used topically and is available over-the-counter to treat sunburns, hence the brand name **Solarcaine.** Injectable and patch forms of **lidocaine** are also available by prescription. Paramedics often use **lidocaine** for arrhythmias in emergencies as an injectable, which is part of the L-E-A-N acronym for the emergency medicines: **lidocaine, epinephrine, atropine**, and **naloxone**.

### ANTIVERTIGO

**Meclizine (Dramamine)**
*MECK-luh-zeen (DRAH-mah-mean)*

You can also see "i-z-i-n" from dizziness in the generic name **meclizine**. **Antivert** is another brand name of the antivertigo drug **meclizine**.

## II. SEDATIVE-HYPNOTICS (SLEEPING PILLS)

A patient came into the pharmacy saying his wife wanted **Tylenol PM**, but saw how expensive the brand name product was. I told him that **Tylenol PM** contains **acetaminophen** and **diphenhydramine** and that he could buy both generics separately and it could be cheaper. He thought about it a minute and said, "It might be a little cheaper at first, but when my wife sends me back to get **Tylenol PM**, what she asked for, it might not be cheaper after all." This story highlights the importance of reading OTC labels closely. It's an opportunity for practitioners to help patients understand OTC products.

Sedative-hypnotics like **diphenhydramine** help patients sleep. Most prescription sedative-hypnotics provide hints about their function in their brand names: **Eszopiclone (Lunesta)** contains Luna for "moon," **ramelteon (Rozerem)** refers to REM sleep, and **zolpidem (Ambien)** creates an "ambient" (tranquil) environment.

Although benzodiazepines such as **clonazepam (Klonopin), diazepam (Valium),** and **lorazepam (Ativan)** work as sedative hypnotics, I discuss benzos later in this chapter as they have other functions as well.

## OTC - Non-narcotic analgesic / Sedative-hypnotic

**Acetaminophen/Diphenhydramine (Tylenol PM)**
*uh-seat-uh-MIN-no-fin / dye-fen-HIGH-dra-mean (TIE-len-all pee em)*

It's common to combine two drugs like **acetaminophen** for aches and **diphenhydramine,** a sedating 1st- generation antihistamine. One of my students came up with take both "phens" when you want to end up sleepin'.

## Benzodiazepine-like

**Eszopiclone (Lunesta)**
*es-zo-PEH-clone (Lou-NES-tuh)*

**Eszopiclone's** generic stem "-clone" will put you in the sleeping zone. Some students point to the "z" in **eszopiclone** for getting your z's. The brand name **Lunesta** uses the Latin for moon (Luna) plus part of the word "rest," which makes it memorable.

**Zolpidem (Ambien, Ambien CR)**
*zole-PEH-dem (AM-bee-en)*

Use the "-pidem" stem to remember **zolpidem** as a sedative-hypnotic. Some students match the brand name **Ambien** with an ambient, sleepy environment. Controlled Release **zolpidem, Ambien CR**, works for people who have difficulty maintaining sleep (DMS) *and* those who have difficulty falling asleep (DFA). The regular version only works for those with DFA.

### MELATONIN RECEPTOR AGONIST

**Ramelteon (Rozerem)**
*ra-MEL-tee-on (row-ZER-em)*

The "–melteon" stem in **ramelteon** lets you know it's a melatonin agonist. Another student said the "m-e-l" in **ramelteon** reminded her of mellow. In **Rozerem**, you can see the "z" for z's, the "r-e-m" for REM (rapid eye movement) sleep. Also, "roze" rhymes with doze.

## III. ANTIDEPRESSANTS

Antidepressant classes often intimidate students, so let's take them one word at a time. The selective serotonin reuptake inhibitors (SSRIs) class includes drugs such as **citalopram (Celexa), escitalopram (Lexapro), sertraline (Zoloft), paroxetine (Paxil)**, and **fluoxetine (Prozac)**. These medications will selectively inhibit reuptake (breakdown) of serotonin within neurons. Increased serotonin levels can improve mood. Note: **escitalopram** has the same "es" prefix (added onto **citalopram**) discussed with the PPIs **esomeprazole** and **omeprazole** where the sinister "S" form is superior.

Similar to the SSRIs are the serotonin-norepinephrine reuptake inhibitors (SNRIs) **duloxetine (Cymbalta)** and

**venlafaxine (Effexor)**. Be careful - **duloxetine** is an SNRI, yet has the "-oxetine" stem of some SSRIs **(fluoxetine, paroxetine)**.

SSRIs and SNRIs carry the names of the neurotransmitters they affect. The tricyclic antidepressant (TCA) class name comes from the chemical structure's three rings. **Amitriptyline (Elavil)** is an example.

The last group of antidepressants includes the monoamine oxidase inhibitors (MAOIs). A word that ends in "ase" is usually an enzyme, so if an antidepressant blocks the enzyme that breaks a neurotransmitter down, then there is more neurotransmitter (the monoamines, in this case) available to elevate the patient's mood. An example of an MAOI is **isocarboxazid (Marplan)**.

## SELECTIVE SEROTONIN REUPTAKE INHIBITORS (SSRIs)

**Citalopram (Celexa)**
*si-TAL-oh-pram (sell-EX-uh)*

> Most students, when seeing two drugs with the same root, **citalopram** and **escitalopram**, quickly put them into long-term memory. One student's trick was to remember that the "p-r-a-m" medications are for de-p-r-ession, but "pram" is not an official stem. I associate the brand name **Celexa** with the word "relax."

**Escitalopram (Lexapro)**
*es-si-TAL-oh-pram (LECKS-uh-pro)*

> **Lexapro** takes the last four letters of **Celexa** and adds "pro." You can think of this as the pro-fessional upgrade, as **Lexapro** came after **Celexa**. It's

common for an S isomer to come to market after the original drug has become available as a generic.

**Ser<u>traline</u> (Zoloft)**
*SIR-tra-lean" (ZO-loft)*

One should use the "-traline" stem to remember **ser<u>traline</u>** is an SSRI, but most students also memorize the brand name **Zoloft** as "<u>loft</u>ing" a depressed patient's mood.

**Flu<u>oxetine</u> (Prozac, Sarafem)**
*flue-OX-uh-teen (PRO-Zack)*

**Fluoxetine** was the first SSRI to make it to market. The "–oxetine" ending is supposed to be for **fluoxetine**-like entities, but you will see "-oxetine" on the SNRI **dul<u>oxetine</u>** (**Cymbalta**) and ADHD medication **atom<u>oxetine</u>** (**Strattera**), so be careful.

When **flu<u>oxetine</u>** gained a new indication, for premenstrual dysphoric disorder (PMDD), it also gained a new brand name: **Sarafem** – "Sara" like the girl's name and "fem" for feminine. The highest ranked angels are Sera-p-h-i-m, so combatting PMDD is the work of angels.

I don't know if that's what the drug manufacturer was going for. In addition, by taking a new brand name for another indication, it might have prevented the potential confusion of a patient with depression on **Prozac** and a patient with PMDD on **Sarafem**. The combination of "pro" for positive and the strong sounding "zac" ending makes **Prozac** sound like a strong antidepressant.

**Paroxetine (Paxil, Paxil CR)**
*par-OX-eh-teen (PACKS-ill)*

**Paroxetine** is similar to the SSRI **fluoxetine** with the same "-oxetine" stem. **Paxil** takes "p-a-x-"" from **pa**ro**x**etine. The controlled-release CR version of **Paxil** is supposed to have fewer initial side effects and be a little easier to dose.

## SEROTONIN-NOREPINEPHRINE REUPTAKE INHIBITORS (SNRIs)

**Duloxetine (Cymbalta)**
*doo-LOX-eh-teen (SIM-bal-tah)*

**Duloxetine** affects serotonin and norepinephrine and one can think of the "du" as duo (two). I have never seen an unhappy cymbal player in a band and "alta" means tall. Students can use either mnemonic to remember **Cymbalta** elevates mood.

**Venlafaxine (Effexor)**
*ven-luh-FAX-een (Eff-ECKS-or)*

This drug is best memorized by its stem "-faxine." If you look at the "afax" in **venlafaxine** and "Effex" in **Effexor,** you can see some commonalities. **Desvenlafaxine** (**Pristiq**), in the "Memorizing 350 Drugs" chapter is an SNRI also.

## TRICYCLIC ANTIDEPRESSANTS (TCAs)

**Amitriptyline (Elavil)**
*ah-meh-TRIP-ta-lean (ELLE-uh-vill)*

Using the "tri" in **amitriptyline** helps students remember this is a "T-C-A," or tricyclic antidepressant. It "trips" up depression. Think of the brand **Elavil** as elevating the patient's mood.

## MONOAMINE OXIDASE INHIBITOR (MAOI)

**Isocarboxazid (Marplan)**
*iso-car-BOX-uh-zid (MAR-plan)*

A student came up with **Marplan** for the atypical sad man who laments, "I so carve boxes" for **isocarboxazid**. In addition, you can take the "m" and "a" from **Marplan** to remember it's an M-A-O-I.

# IV. SMOKING CESSATION

There are many nicotine replacement products on the market. I've chosen to focus on two tablets used by prescription that help patients quit smoking: **bupropion (Wellbutrin, Zyban)** and **varenicline (Chantix)**.

**Bupropion (Wellbutrin, Zyban)**
*byoo-PRO-pee-on (well-BYOO-trin, ZY-ban)*

**Bupropion** was first marketed as **Wellbutrin**, an atypical antidepressant, one that didn't fit into the SSRIs, SNRIs, TCAs, or MAOIs. Reports must have come in that patients stopped smoking while taking it, so the company repackaged the drug with a new brand name as **Zyban**, to put a "ban" on smoking. There is some risk with this medication, especially in patients with a history of seizures.

**Varenicline (Chantix)**
*Vah-WREN-eh-clean (CHAN-ticks)*

**Varenicline**, another smoking cessation medication, has caused distressing dreams, suicidal thoughts, and other adverse effects. A student, in a southern drawl, remembered varenicline's generic name by saying, "With **varenicline**, 'I'm vary incline ta quit.'" Another came up with "My chant is, 'I don't need my fix' with **Chantix.**"

## V. Benzodiazepines

Benzodiazepines relieve anxiety, insomnia and muscle spasms. Like the tricyclic antidepressants, benzodiazepines get their name from their chemical structure: a benzene ring and a diazepine ring. Because benzodiazepine has many syllables, most people call them benzos. Examples include **alprazolam (Xanax), clonazepam (Klonopin), diazepam (Valium)**, and **lorazepam (Ativan)**. Note benzodiazepines have the similar generic suffixes "-azolam" and "-azepam." Benzos replaced barbiturates as a sleep aid because barbiturates can cause respiratory depression. A student remembered this by thinking of barbiturates (barbs) as literal barbs on razor wire fences that puncture lungs.

**Alprazolam (Xanax)**
*Al-PRAY-zo-lam (ZAN-ax)*

You should remember benzodiazepines by their two endings, "-azolam" or "-azepam." Be careful as a number of online and highly regarded test prep resources, including those by licensed professionals, refer to benzodiazepine stems as "-pam" or "-lam." That is incorrect. This will lead you to think drugs that are not benzos are in the class. For example, **verapamil (Calan)** is a calcium channel blocker and **lamotrigine (Lamictal)** is an antiepileptic. I've even seen a well-regarded resource indicate that **citalopram** has a -pam suffix at the end, and that's understandable as some patients drop the "r," pronouncing it "citalo-pam," but this is incorrect. **Alprazolam** has one z; benzodiazepine has two; **Xanax** sounds like a "z" to help you get a snooze and "x's" out anxiety too. You can also look at the word **Xanax** and see the "a-n-x" from anxiety.

**Midazolam (Versed)**
*meh-DAZE-oh-lam (VER-said)*

You should remember **midazolam** through the "-azolam" stem, but there are other tips you can use. There are two m's in **midazolam** for the memories you can't form since **midazolam** causes anterograde amnesia. Just as your *ante*brachium is your forearm, and the *ante* is the money poker players put out before the dealer deals, *ante*rograde amnesia is the inability to form memories. Alternatively, you can use the brand name; I can't remember the "verse you just said" for **Versed**.

**Clonazepam (Klonopin)**
*kloe-NAZ-uh-pam (KLON-uh-pin)*

**Clonazepam** should be remembered from the "-azepam" stem. The brand name **Klonopin** uses the phonetic spelling of generic **clonazepam's** first four letters "c-l-o-n."

**Lorazepam (Ativan)**
*lore-A-zeh-pam (AT-eh-van)*

Remember **lorazepam** through the "-azepam" stem or think about the brand name **Ativan** vanquishing anxiety.

## VI. ADHD MEDICATIONS

**ADHD** (attention-deficit-hyperactivity disorder) stimulants calm a patient who has a hyperactive mind and / or body without a sedative effect. Examples include the drugs **dexmethylphenidate (Focalin)** and **methylphenidate (Concerta)**. These two medications have the same root, **methylphenidate**. In chemistry, compounds direct plane-polarized light to either the left or right. These terms are "d" or "(+)" for dextrorotatory compounds rotating plane-polarized light to the right, and "l" or "(-)" for levorotatory compounds rotating plane-polarized light to the left. **Dexmethylphenidate** rotates plane-polarized light to the right. **Dexmethylphenidate** should be more effective, last longer, and have fewer side effects than methylphenidate.

**Atomoxetine (Strattera)** is a non-stimulant medication and because there's not a potential for abuse, it doesn't carry a DEA schedule. It's *not* an SSRI like **fluoxetine**, even though it ends in "-oxetine."

### STIMULANT – SCHEDULE II

**Dexmethylphenidate (Focalin)**
*dex-meth-ill-FEN-eh-date (FOE-ca-lin)*

When I was a student, I could remember that **Focalin** and **Concerta** both had **methylphenidate** in their names, but I could never remember which was

**dexmethylphenidate** and which was **methylphenidate**. Then I thought of the "F" in "**Focalin**" as following the "d-e" (like in the alphabet) in **dexmethylphenidate** to help me. To remember **Focalin** is for ADHD, you think of **Focalin** helping a patient focus.

**Methylphenidate (Ritalin, Concerta)**
*meth-ill-FEN-eh-date (con-CERT-uh)*

**Methylphenidate** has many brand names, including **Ritalin** and **Concerta.** Most students seem to already know **methylphenidate,** but remember that **Concerta** can help a patient concentrate. To remember it's an amphetamine, one student said she thought of staying up all night at a concert with **Concerta. Concerta** is a long-acting medication that needs to be taken only once a day.

### NON-STIMULANT – NON-SCHEDULED

**Atomoxetine (Strattera)**
*AY-toh-mocks-e-teen (stra-TER-uh)*

Note the "–oxetine" stem here is also not an SSRI antidepressant, but a non-stimulant medication for ADHD. **Strattera** helps straighten patients' attention.

## VII. BIPOLAR DISORDER

Mood stabilizers such as **lithium** are especially likely to cause electrolyte imbalances. If you look at the periodic table, you see that lithium (Li) and sodium (Na) are in the same group (the alkali metals) and both have a +1 charge as an ion. Because of this similarity, what happens to sodium will happen to **lithium,** causing either a toxic or a subtherapeutic state if too much **lithium** is retained or excreted, respectively. Other meds, like **risperidone**, can help control certain symptoms of the disease until the **lithium** level is correct.

### SIMPLE SALT

**Lithium (Lithobid)**
*LITH-e-um (LITH-oh-bid)*

> **Lithium** sits in the same group on the periodic table of elements as the Latin *Natrium* (Na), commonly known as the chemical element sodium. The body has trouble telling the difference between lithium and sodium, and too much or too little salt can wreak havoc on **lithium** levels. A way to remember this is the saying, "Where the salt goeth, the **lithium** goeth." The "b-i-d" in the brand name **Lithobid** is the Latin *bis-in-die*, or a twice-daily dosing schedule.

## VIII. SCHIZOPHRENIA

We break the medication classification for schizophrenia into typical (1st-generation) or atypical (2nd-generation). We further divide the typical antipsychotics **chlorpromazine (Thorazine)** and **Haloperidol (Haldol)** into low potency and high potency, respectively.

While these two antipsychotics have the same therapeutic effects, their side effect profiles are different. Low potency drugs like **chlorpromazine** cause more sedation, but fewer extrapyramidal symptoms. Extrapyramidal symptoms are movement disorders associated with antipsychotics.

High potency drugs like **haloperidol** cause more extrapyramidal symptoms (movement disorders), but less sedation. We prescribe typical antipsychotics for positive symptoms such as delusions, hallucinations and paranoia.

Atypical antipsychotics such as **quetiapine (Seroquel)** and **risperidone (Risperdal)** cause fewer extrapyramidal effects, but have more negative metabolic effects like weight gain, diabetes, and hyperlipidemia. Second-generation drugs work for positive symptoms, such as delusions, as well as negative symptoms, such as poor motivation and emotional and social withdrawal.

## FIRST GENERATION ANTIPSYCHOTIC (FGA) LOW POTENCY

**Chlorpromazine (Thorazine)**
*Klor-PRO-mah-zeen (THOR-uh-zeen)*

> **Chlorpromazine** was the first antipsychotic. While it carries side effects, it represented a new treatment option for schizophrenic patients. Generational classifications are especially important in antipsychotics because of differences in both side effects and effects on positive versus negative symptoms. Thor is a mythical god, and you can think of **Thorazine** as helping people who have delusions of mythical people.

## FIRST GENERATION ANTIPSYCHOTIC (FGA) HIGH POTENCY

**Haloperidol (Haldol)**
*hal-low-PEAR-eh-doll (HAL-doll)*

> Use the "-peridol" stem to recognize this 1st-generation high potency drug. Many students think

of the "halo" in **haloperidol** to remember this is high potency. To make the brand name **Haldol**, they just took the first and last three letters of the generic name **haloperidol**.

### SECOND-GENERATION ANTIPSYCHOTICS (SGA) (ATYPICAL ANTIPSYCHOTICS)

**Risperidone (Risperdal)**
*ris-PEAR-eh-done (RIS-per-doll)*

Note the stem "-peridol" from **haloperidol** and "-peridone" from **risperidone** are similar; that can help you remember both of these are antipsychotics. The brand name **Risperdal** and generic name **risperidone** share the opening letters "r-i-s-p-e-r." One student thought of risper and whisper, as in calming the whispering voices.

**Quetiapine (Seroquel)**
*Kweh-TIE-uh-peen (SEAR-uh-kwell)*

In **quetiapine**, you should use the "-tiapine" stem to recognize that this is a 2nd-generation antipsychotic. If you switch the "i" and "t" in **quetiapine**, you get the word "quiet," as in quieting the voices. **Seroquel**, the brand name, shares the "q-u-e" from **quetiapine** and quell means to silence someone.

## IX. ANTIEPILEPTICS

The traditional anti-epileptic drugs **carbamazepine (Tegretol)**, **divalproex (Depakote)**, and **phenytoin (Dilantin)** have been around for a long time and we usually know what to expect with their use.

We may have less experience with the newer anti-epileptics, such as **gabapentin (Neurontin)** and **pregabalin (Lyrica)**, but they are often just as effective as the traditional drugs. Neurologists try medications in a patient until one drug relieves the seizure symptoms.

## TRADITIONAL ANTIEPILEPTICS

**Carbamazepine (Tegretol)**
*car-bah-MAZE-uh-peen (TEG-reh-tawl)*

While **carbamazepine's** "-pine," pronounced, "peen," is the stem, it's not very useful because the – "-pine" ending just means a chemical has three rings. Instead, think either of seizures being "carbed" (curbed) or of being "amazed" that the seizures are "stopping" using the letters inside **carb-amaze-pine**. I remember the brand name **Tegretol** has the "t-r-o-l" in control, as in to control seizures.

**Divalproex (Depakote)**
*dye-VAL-pro-ex (DEP-uh-coat)*

While you can find "v-a-l" in many medication names, it is helpful to think of the "val" in **divalproex** and it's similarity with the "v-u-l" in convulsions. I know it's a stretch. One student thought of **divalproex** as a pro at extracting seizures.

**Phenytoin (Dilantin)**
*FEN-eh-toyn (DYE-lan-tin)*

The "–toin" stem helps you remember **phenytoin** is an antiepileptic. **Dilantin** and shakin' almost rhyme.

### Newer Antiepileptics

**Gabapentin (Neurontin)**
*GA-ba-PEN-tin (NER-on-tin)*

The "-gab" stem is a little misleading. Neither **gabapentin** nor **pregabalin** directly affect gamma-amino-butyric-acid (GABA) receptors. However, having them both have the same stem helps set a memory device for the newer antiepileptics. The "neu" in **Neurontin** is one way to remember that this is a newer drug.

**Pregabalin (Lyrica)**
*pre-GAB-uh-lin (LEER-eh-ca)*

A lyre is a musical instrument and a lyric is a line in a song. Either way, you can think of a seizure coming back into harmony with **Lyrica**.

## X. Parkinson's, Alzheimer's, Motion Sickness

Most people associate Parkinson's with the celebrity actor Michael J. Fox of *Back to the Future* fame. Using a drug like **levodopa / carbidopa (Sinemet)** for Parkinson's works to restore dopamine, a neurotransmitter responsible for proper motor function that is seriously depleted by the disease.

Concerning Alzheimer's, I remember vividly calling my grandmother and telling her "I love you," to which she responded, "I hope your wife doesn't find out." There is a cruelty in Alzheimer's, and the patients and their caregivers desperately need your help. A drug like **donepezil** works to restore the neurotransmitter acetylcholine, by reducing its breakdown by acetylcholinesterase.

## PARKINSON'S

**Levodopa / Carbidopa (Sinemet)**
*LEE-vo-doe-pa / CAR-bid-oh-pa" (SIN-uh-met)*

Increased dopamine is critical to dealing with the disease. **Sinemet** combines **levodopa** and **carbidopa** to work synergistically. **Carbidopa** doesn't actually have an antiparkinsonian effect, but it reduces the degradation of **levodopa** so more is available to the patient from a smaller dose.

**Selegiline (Eldepryl)**
*se-LEDGE-eh-lean (EL-duh-pril)*

The "-giline" stem should be your hint that **selegiline** is a Parkinson's medication. You can find the letters of the word senile in the generic name **selegiline**. I used to confuse this as an Alzheimer's medication, but it's not, so I was being a little senile. The brand name **Eldepryl** helps you remember that it relieves symptoms of Parkinson's disease, a condition more prevalent in the "elderly."

## ALZHEIMER'S

**Donepezil (Aricept)**
*Doe-NEP-eh-zill (AIR-eh-sept)*

When I think of **donepezil** as an Alzheimer's medication, I think, "I don't remember zilch!" The brand name **Aricept** improves perception and

Alzheimer's patients' powers of recollection. A student thought of the "air" in **Aricept** as helping a patient who acts somewhat spacey or air-headed.

**Memantine (Namenda)**
*Meh-MAN-teen (Nuh-MEN-duh)*

The generic name **memantine** has "mem" for memory in it. The brand name **Namenda** comes from N-Methyl-D-aspartate **(NMDA)**, the receptor it antagonizes. One of my students always thought of **Namenda** as mending the brain.

## MOTION SICKNESS

**Scopolamine (Transderm-Scop)**
*sco-POL-uh-mean (trans-DERM SCOPE)*

**Transderm-Scop** is a transdermal form of **scopolamine**, for motion sickness. "Trans" means across, "derm" means "skin," so across-the-skin **scopolamine**. It's often used for patients on cruise ships.

## Nervous System Drug Quiz (Level 1)

Classify these drugs by placing the corresponding drug class letter next to each medication. Try to underline the stems before you start and think about the brand name and function of each drug.

1. Alprazolam (Xanax)
2. Amitriptyline (Elavil)
3. Atomoxetine (Strattera)
4. Citalopram (Celexa)
5. Dexmethylphenidate (Focalin)
6. Divalproex (Depakote)
7. Haloperidol (Haldol)
8. Isocarboxazid (Marplan)
9. Levodopa / Carbidopa (Sinemet)
10. Zolpidem (Ambien)

**Nervous System Drug Classes:**

A. ADHD drug / non-stimulant
B. ADHD drug / stimulant
C. Antidepressant: MAOI
D. Antidepressant: SNRI
E. Antidepressant: SSRI
F. Antidepressant: TCA
G. Antiepileptic: Newer
H. Antiepileptic: Traditional
I. Antipsychotic: Atypical
J. Antipsychotic: Typical
K. Benzodiazepine
L. Parkinson's
M. Sedative-hypnotic
N. Simple salt

## NERVOUS SYSTEM DRUG QUIZ (LEVEL 2)

Classify these drugs by placing the corresponding drug class letter next to each medication. Try to underline the stems before you start and remember the brand name and function of each drug.

1. Venlafaxine
2. Diazepam
3. Phenytoin
4. Quetiapine
5. Isocarboxazid
6. Eszopiclone
7. Fluoxetine
8. Citalopram
9. Selegiline
10. Lithium

**Nervous System Drug Classes:**

A. ADHD drug/non-stimulant
B. ADHD drug/stimulant
C. Antidepressant: MAOI
D. Antidepressant: SNRI
E. Antidepressant: SSRI
F. Antidepressant: TCA
G. Antiepileptic: Newer
H. Antiepileptic: Traditional
I. Antipsychotic: Atypical
J. Antipsychotic: Typical
K. Benzodiazepine
L. Parkinson's
M. Sedative-hypnotic
N. Simple salt

## NEURO: MEMORIZING THE CHAPTER

| | | |
|---|---|---|
| **Benzocaine** | Amitriptyline | Chlorpromazine |
| **Lidocaine** | Isocarboxazid | Haloperidol |
| **Meclizine** | Bupropion | Risperidone |
| **Acetaminophen PM** | Varenicline | Quetiapine |
| Eszopiclone | Alprazolam | Carbamazepine |
| Zolpidem | Midazolam | Divalproex |
| Ramelteon | Clonazepam | Phenytoin |
| Citalopram | Lorazepam | Gabapentin |
| Escitalopram | Dexmethylphenidate | Pregabalin |
| Sertraline | Methylphenidate | Levodopa/carbidopa |
| Fluoxetine | Atomoxetine | Selegiline |
| Paroxetine | Lithium | Memantine |
| Duloxetine | | Donepezil |
| Venlafaxine | | Scopolamine |

### "N" NEURO

The four neuro OTC drugs are the local anesthetics **benzocaine** and **lidocaine**, which are an ester and amide respectively. **Meclizine** is for dizziness, and **acetaminophen / diphenhydramine** is for insomnia. Connect this with other anti-insomniacs the benzodiazepine-like sedative-hypnotics **eszopiclone**, **zolpidem**, and melatonin receptor agonist **ramelteon**. No sleep = depression. Five SSRI antidepressants: **citalopram** and **escitalopram**, **sertraline** in the middle followed by three "-oxetines:" **fluoxetine** and **paroxetine** (the SSRIs) and **duloxetine** (the SNRI) and **venlafaxine** the SNRI; followed by, in order of safety, the TCA **amitriptyline** and the MAOI **isocarboxazid**. Move from depression to smoking: **bupropion** and **varenicline**, smoking while anxious, two –azolams: **alprazolam** and **midazolam** and two -azepams **clonazepam** and **lorazepam**. "A" anxious to the "A"DHD stimulants **dexmethylphenidate, methylphenidate**, to non-stimulant **atomoxetine**; to "m-o" from atomoxetine for m-o-o-d stabilizer **lithium**; then "L" from lithium to low potency 1st-generation **chlorpromazine**, to high potency "halo" **haloperidol**, "–peridol" to "–peridone," 2nd generation whisper **risperidone** to whisper-quiet **quetiapine**; pine to traditional antiepileptics **carbamazepine**, **divalproex** and **phenytoin** to two newer antiepileptics Neu-rontin (which is **gabapentin**) and **pregabalin**. From epileptic motion to Parkinsonian motion: **carbidopa / levodopa** and **selegiline** senility to Alzheimer's memory is done: **memantine** to "D" **donepezil** and "D" dizzy **scopolamine**.

# CHAPTER 6 CARDIO

## I. OTC ANTIHYPERLIPIDEMICS AND ANTIPLATELET

Few over-the-counter (OTC) medications help a patient with cardiologic issues. Both **Omega-3-acid ethyl esters (Lovaza)** and **Niacin (Niaspan ER)** come as brand name and OTC products, and are used to reduce cholesterol's impact on the patient. Plain **aspirin** (**Ecotrin**) in a low-dose of 81 milligrams helps prevent platelets from clotting, reducing a patient's chance of a heart attack.

### OTC ANTIHYPERLIPIDEMICS

**Omega-3-acid ethyl esters (Lovaza)**
*Oh-MEG-uh THREE AS-sid ETH-ill EST-ers (Loh-VAH-zah)*

> **Omega-3-fatty acids** are available over-the-counter, but there is also a prescription version that undergoes rigorous FDA testing. **Lovaza** is a prescription brand name, but you can find **omega-3-fatty acids** over-the-counter commonly labeled as "Fish Oil."

**Niacin (Niaspan ER)**
*NYE-uh-sin (NYE-uh-span ee-ar)*

> A vitamin like **niacin** can reduce cholesterol levels in the body, however it may cause facial flushing that an **aspirin** thirty minutes before treatment prevents.

### OTC Antiplatelet

**Aspirin (Ecotrin)**
*AS-per-in (ECK-oh-trin)*

> The 81mg daily **aspirin** dosage is not for analgesia or fever reduction. Rather, it keeps platelets from sticking, helping prevent strokes and heart attacks.

## II. Diuretics

The order of important structures in the nephron goes from *glomerulus* to *proximal convoluted tubule* (something in close proximity is near) to *Loop of Henle* to *distal convoluted tubule* (something that's distant or distal is far) to the *collecting duct*. The order of diuretics would then be:

1. Osmotic diuretics like **mannitol** (**Osmitrol**) work at the proximal convoluted tubule (PCT).

2. Loop diuretics like **furosemide** (**Lasix**) affect the Loop of Henle.

3. Thiazide diuretics like **hydrochlorothiazide** (**Microzide**) work at the distal convoluted tubule (DCT).

4. Potassium sparing diuretics like **triamterene** (**Dyrenium**) and **spironolactone** (**Aldactone**) work at the collecting duct.

Picture a water slide. A lot of water flows at the top (the glomerulus). A trickle flows at the bottom (the collecting duct). Similarly, diuretics produce less diuresis as they continue down the waterslide. The order from most to least diuresis is osmotic > loop > thiazide > potassium sparing.

## OSMOTIC

**Mannitol (Osmitrol)**
*MAN-eh-tall (OZ-meh-trawl)*

**Mannitol**, an osmotic diuretic reduces intracranial pressure in an emergency. The brand name **Osmitrol** combines the class of medication "osmotic," and adds that it helps control brain swelling. The actor Bruce Lee died from this event.

## LOOP

**Furosemide (Lasix)**
*Fyoor-OH-seh-mide (LAY-six)*

Chemists named this class of diuretics after the part of the nephron the drug works in, the Loop of Henle. While the "–semide" stem indicates a "furosemide-type" diuretic, that's like defining a word with the word itself. One student said, "I have to pee furiously" as her mnemonic since loop diuretics produce significant diuresis. The brand name **Lasix** indicates it lasts six hours.

## THIAZIDE

**Hydrochlorothiazide (Microzide)**
*High-droe-klor-oh-THIGH-uh-zide (MY-crow-zide)*

Thiazide diuretics get their class name from the stem of generic drugs like **hydrochlorothiazide**. The abbreviation HCTZ comes from "h" for hydro, "c" for chloro, "t" for thia, and "z" for zide. Thiazides don't produce as much diuresis as loop diuretics, but are excellent for initial treatment of hypertension. While the "hydro" in **hydrochlorothiazide** stands for the hydrogen atom, you can think of "hydro" as "water" for diuretic. The brand **Microzide** has thiazide's last four letters.

## POTASSIUM SPARING AND THIAZIDE

**Triamterene / Hydrochlorothiazide (Dyazide)**
*try-AM-terr-een / High-droe-klor-oh-THIGH-uh-zide (DIE-uh-zyde)*

The combination of a potassium sparing diuretic (**triamterene**) and thiazide (**hydrochlorothiazide**) keeps potassium levels in balance while producing modest diuresis. The brand name **Dyazide** is triamterene's old brand name **Dyrenium** plus the last five letters of **hydrochlorothiazide**.

## POTASSIUM SPARING

**Spironolactone (Aldactone)**
*spear-oh-no-LACK-tone (Al-DAK-tone)*

**Spironolactone** is another potassium sparing diuretic, but this drug can cause gynecomastia. Gynecomastia is an enlargement of male breasts. To remember **spironolactone** works in the collecting duct, I look at the "lactone," and think "last one." I know a lactone is a kind of chemical structure, but its place of action sticks in my head with this mnemonic.

To come up with the brand name, the manufacturer simply replaced the "spironol" of **spironolactone** with "ald" to make **Aldactone**. The "ald" is especially important because **spironolactone** blocks aldosterone, an important steroid hormone that helps the body retain sodium and water when blood pressure drops.

### ELECTROLYTE REPLENISHMENT

**Potassium chloride (K-Dur)**
*poe-TASS-ee-um klor-eyed (Kay-Dur)*

**Potassium chloride** is a supplement often administered when a potassium sparing diuretic is contraindicated or when a loop diuretic lowers a patient's potassium levels. The "K" in **K-Dur** is the chemical symbol for potassium. The "Dur" is for long duration.

## III. UNDERSTANDING THE ALPHAS AND BETAS

Confusion about alpha-adrenergic antagonists like **doxazosin (Cardura)** and beta-adrenergic antagonists like **propranolol (Inderal)** comes from seeing the receptor names, alpha and beta, instead of drug classifications; e.g., both **doxazosin** and **propranolol** are blood pressure lowering pills.

Alpha and beta are the first two letters of the Greek alphabet and name the receptors where these medications work. An adrenergic *agonist* works *like* adrenaline while an adrenergic *antagonist* works in the *opposite* way.

The prefix "adren" refers to the adrenal glands. The adrenal glands are *above* (ad) the *kidney* (renal) and secrete adrenaline. The suffix "ergic" refers to the Greek for "works like." Therefore, these drugs work like **adrenaline**. Note: **Adrenaline** and **epinephrine** are the same. **Epinephrine** uses the Greek translation of *above* (epi) and *kidney* (neph) to make **epinephrine** instead of the Latin form, **adrenaline**.

Instead of calling a drug a blood pressure pill (antihypertensive), its therapeutic class, prescribers classify a drug by the receptor it affects. By calling **propranolol** (**Inderal**) a beta-blocker, it's easier not to pigeonhole a drug into one use. Beta-blockers, for example, can treat angina pectoris, congestive heart failure, stage fright, and migraine, in addition to hypertension. **Doxazosin**, the alpha-blocker, also has multiple uses, including hypertension and benign prostatic hyperplasia (BPH). This is why classifying by the receptor name alpha or beta makes more sense.

Furthermore, there are receptor sub-types. Beta-1 receptors are concentrated in the heart (and we have one heart), and beta-2 receptors are concentrated in the lungs (and we have two lungs). You can find them in other places in the body, but for our introductory purposes, it's useful to think in this way.

If a beta-blocker is *non-selective*, like **propranolol (Inderal)**, it can affect both beta-1 receptors in the heart to lower heart rate (good), block beta-2 receptors in the lungs and cause bronchoconstriction (bad). An asthmatic patient might have an adverse reaction to a drug that bronchoconstricts as a side effect.

We prefer **metoprolol (Lopressor)** to control blood pressure because it's selective for just the heart; it's classified as *beta-1 selective*. However, the body will try to compensate for this reduction in blood pressure by vasoconstricting arterioles.

**Carvedilol (Coreg)**, a 3rd-generation beta-blocker, shows that it might be the best choice because it has vasodilating effects to counteract the vasoconstriction as well as cardiac effects.

## ALPHA-1 ANTAGONIST

**Doxazosin (Cardura)**
*Docks-AZ-oh-sin (car-DUR-uh)*

The blockade of alpha-1 receptors by **doxazosin** causes vasodilation and subsequent reduction in blood pressure. Memorize the stem "–azosin" as an alpha-blocker. You can also link the brand name, as **Cardura** provides durable cardiac relief of hypertension.

## ALPHA-2 AGONIST

**Clonidine (Catapres)**
*KLAH-neh-deen (CAT-uh-press)*

**Clonidine** works in the brain by affecting alpha-2 receptors to reduce peripheral vascular resistance. You can look at the brand name **Catapres** and think of catabolize (break down) pressure (blood pressure).

Prescribers use **clonidine** in ADHD as single therapy or with stimulants like **methylphenidate (Concerta)**. I had the weirdest experience at the gym. A parent had a loud and lengthy discussion with a psychiatrist about her child's **clonidine** and **Concerta** while lifting weights. I never forgot the **clonidine / Concerta** tandem after that.

## Beta blockers – 1st-generation – non-beta-selective

**Propranolol (Inderal)**
*Pro-PRAN-uh-lawl (IN-dur-all)*

I had a student tell me about hearing a British pronunciation of beta as "bee-tah" and she thought, "Oh, laugh out loud" in a British accent to remember the "-olol" stem in **propranolol**. If you think of the last "al" in the brand name **Inderal** as it blocks "all" beta-receptors, you can remember this is a non-selective blocker.

## Beta blockers – 2nd-generation – beta-selective

**Atenolol (Tenormin)**
*uh-TEN-oh-lol (Teh-NOR-min)*

While the "-olol" in **atenolol** identifies this medication as a beta-blocker, a student does have to memorize that **atenolol** is 2nd generation. You can do that by memorizing its position after a non-selective first generation **propranolol** in this book or by seeing the "ten" in **atenolol** and knowing it's divisible by two. The "Ten" in the brand name **Tenormin** also matches to the "ten" in atenolol.

**Metoprolol tartrate (Lopressor)**
*meh-TOE-pruh-lawl TAR-trait (low-PRESS-or)*

Practitioners rarely highlight the distinction in salts like tartrate and succinate, but it's important to recognize, as **metoprolol tartrate** and **metoprolol succinate** work for different lengths of time.

It would have been nice if the succinate was short acting and the tartrate, long acting; then alphabetical order would have worked or "s" for short acting. That it goes contrary to this logic is how I remember which is which. You can use the brand name **Lopressor** to remind you that **Lopressor** lowers blood pressure.

**Metoprolol succinate (Toprol XL)**
*meh-TOE-pruh-lawl SUCKS-sin-ate" (TOE-prall ex-ell)*

**Metoprolol succinate** is a long-acting form of **metoprolol**. The XL, often used to identify clothing as extra-large, indicates an extra-long acting effect in medications like **Toprol XL**.

## BETA BLOCKERS – 3RD-GENERATION – NON-BETA-SELECTIVE, VASODILATING

**Carvedilol (Coreg)**
*car-veh-DILL-awl (CO-reg)*

I'm not sure if it was intentional to create a kind of hybrid stem with the "dil" replacing the first "o" in "olol," but you can remember **carvedilol** works by both vasodilation and beta-blockade in this way. The only official stem, however, is the "-dil-." I remember the brand name **Coreg** because it regulates coronary function.

## IV. THE RENIN-ANGIOTENSIN-ALDOSTERONE-SYSTEM DRUGS

The RAAS, or renin-angiotensin-aldosterone system, controls blood pressure. By defining a few words in this system, we can better understand how the drugs work. The word **renin** comes from **renal** for kidneys, and this enzyme converts angiotensinogen to angiotensin I. Angiotensin converting enzyme (ACE) converts **angiotensin I** to **angiotensin II**. Angiotensin II is a potent vasoconstrictor and increases blood pressure when that is what our body needs. **Aldosterone** causes the retention of sodium and water, which can further increase blood pressure.

Angiotensin converting enzyme inhibitors (ACE inhibitors) such as **enalapril (Vasotec)** and **lisinopril (Zestril)** stop the body from creating this potent vasoconstrictor, thereby reducing hypertension.

ARBs, or angiotensin II receptor blockers, such as **losartan (Cozaar), olmesartan (Benicar)**, and **valsartan (Diovan)**, block or inhibit the connection between angiotensin II and the receptor that would cause vasoconstriction. This class of drugs is often used as an alternative to an ACE inhibitor when a patient experiences cough as a side effect from an ACE inhibitor.

Here is a mnemonic a student of mine who loved literature made up. It might help you remember the difference: D'artagnan the musketeer has to be "sartan" with the bARB of his blade, otherwise, he'll not be an ACE in April, I'm afraid.

## Angiotensin converting enzyme inhibitors (ACEIs)

**Enalapril (Vasotec)**
*eh-NAL-uh-pril (VA-zo-teck)*

Sometimes students simply refer to the ACEIs like **enalapril** as "prils" based on the stem "-pril." While **enalapril** is taken orally, **enalaprilat** is an injectable form and active metabolite of **enalapril**. The brand **Vasotec** alludes to vasodilation on the vasculature.

**Lisinopril (Zestril)**
*lie-SIN-oh-pril (ZES-tril)*

**Lisinopril**, like **enalapril**, works to block the vasoconstricting effects of angiotensin II. A student came up with "**Lisinopril** thrills an overworked heart, blocking angiotensin II from getting a start."

## Angiotensin II receptor blockers (ARBs)

**Losartan (Cozaar)**
*low-SAR-tan (CO-zar)*

Angiotensin II receptor blockers like **losartan** are often called ARBs and should be learned by the suffix "-sartan." The brand name **Cozaar** looks like it has R-A-A-S backwards (for renin-angiotensin-aldosterone-system) with a "z" replacing the "s."

**Olmesartan (Benicar)**
*Ole-meh-SAR-tan (BEN-eh-car)*

**Olmesartan** is another ARB identified by its "-sartan" stem. The brand name **Benicar** hints that the drug will benefit the cardiac system.

**Valsartan (Diovan)**
*val-SAR-tan (DYE-oh-van)*

Identify the generic name **valsartan** by its "-sartan" stem. **Diovan** has three of the letters of the generic name valsartan.

## V. Calcium channel blockers (CCBs)

Both calcium channel blocker (CCB) classes, the non-dihydropyridines and dihydropyridines, are vasodilators. However, the non-dihydropyridines **diltiazem (Cardizem)** and **verapamil (Calan)** also affect the heart directly and are antidysrhythmics. **Amlodipine (Norvasc)** and **nifedipine (Procardia)** are two dihydropyridines that only vasodilate.

If a patient needs a calcium channel blocker to prevent uterine contractions, **nifedipine (Procardia)** would be the best choice because it does not suppress the mother's and fetus's hearts as the non-dihydropyridines would.

In our daughters' case, the doctor prescribed low dose **nifedipine** so the calcium channel blockers did not suppress four hearts - my wife's and three unborn daughters'.

### Non-dihydropyridines

**Diltiazem (Cardizem)**
*dill-TIE-uh-zem (CAR-deh-zem)*

The "-tiazem" stem identifies **diltiazem** as a non-dihydropyridine. The brand name **Cardizem** adds the first five letters from cardiac to the last three letters of the generic diltiazem.

**Verapamil (Calan)**
*ver-APP-uh-mill (KALE-en)*

One of my students came up with "Vera and Pam are ill and need this calcium blocking cardiac pill," for **verapamil**. Often **verapamil** is associated with constipation. My grandmother, a Navy nurse, used to put a **verapamil** tablet on my grandfather's breakfast cereal spoon. I always thought my grandfather was silently praying before he ate. When I finally asked him why he was so quiet, he said something to the effect of, "I'm deciding whether I want to eat or poop today." The brand name **Calan** takes three letters from the word calcium and two from channel blocker.

## DIHYDROPYRIDINES

**Amlodipine (Norvasc)**
*am-LOW-duh-peen (NOR-vasc)*

Students usually recognize **amlodipine's** "-dipine" stem, not only as a dihydropyridine, but also as a dip in blood pressure. A way to remember the brand name **Norvasc** is to think of the "n-o-r" from normalizes and the "v-a-s-c" from vasculature.

**Nifedipine (Procardia)**
*nigh-FED-eh-peen (pro-CARD-e-uh)*

**Nifedipine** is a dihydropyridine with the "-dipine" stem. **Procardia** takes the "p-r-o" from "promotes" and "c-a-r-d-i-a" from "cardiac" so you can remember the brand **Procardia** as promoting cardiac health.

## VI. Vasodilator

**Nitroglycerin (Nitrostat)**
*nigh-trow-GLI-sir-in (NYE-trow-stat)*

"Nitro-"is a World Health Organization (WHO) stem. **Nitroglycerin** converts to nitric oxide, a vasodilator. Make sure the patient sits when he takes the med because it causes significant dizziness. With **Nitrostat**, think "nitrous" from sports cars – the patient and blood pressure drop "stat."

## VII. Antihyperlipidemics

Medications for elevated cholesterol fall into several categories, including the "statins," which are more properly called the HMG-CoA reductase inhibitors, and the fibric acid derivatives. It's better to recognize statins such as **atorvastatin (Lipitor)** and **rosuvastatin (Crestor)** with the infix + suffix "–vastatin" because **nystatin (Mycostatin)** an antifungal medication contains "statin" in its name.

### HMG-CoA reductase inhibitors

**Atorvastatin (Lipitor)**
*uh-TORE-va-stat-in (LIP-eh-tore)*

Students use letters of the HMG-CoA class to memorize potential adverse effects: "H" for hepatotoxicity, "M" for myositis, "G" for gestation (can't use during pregnancy). Brand name **Lipitor** is a lipid gladiator.

**Rosuvastatin (Crestor)**
*Row-sue-vuh-STAT-in (CRES-tore)*

Like **atorvastatin, rosuvastatin** shares the "-vastatin" ending. Remember **Crestor** decreases cholesterol.

### FIBRIC ACID DERIVATIVES

**Fenofibrate (Tricor)**
*fen-oh-FIE-brate (TRY-core)*

A drug like **fenofibrate** has the obvious stem "-fibrate," a triglyceride lowering fibric acid derivative. **Tricor** lowers triglycerides to help your coronary status.

## VIII. ANTICOAGULANTS AND ANTIPLATELETS

Anticoagulants affect clotting factors to help prevent thrombosis. The injectable anticoagulants **enoxaparin (Lovenox)** or **heparin** and the oral anticoagulant **warfarin (Coumadin)** affect coagulation in slower moving blood vessels like veins. **Dabigatran (Pradaxa)** works as an anticoagulant, but does not require monitoring with blood tests like **warfarin** and **heparin**.

Platelets stop bleeding by creating clots. However, patients with excess cholesterol might have a plaque that makes the clot more likely in a dangerous place. The antiplatelet drugs **aspirin (Ecotrin)** and **clopidogrel (Plavix)** decrease how "sticky" platelets are in high-pressure vessels such as arteries to prevent the clot and ensuing heart attack or stroke.

## Anticoagulants

**Enoxaparin (Lovenox)**
*e-knocks-uh-PEAR-in (LOW-ven-ox)*

**Enoxaparin** and **heparin** share the "-parin" stem because they are related. **Enoxaparin** is more expensive per dose, but patients can use it at home. It's also used as bridge therapy in a patient who is starting **warfarin** therapy. **Lovenox** is a low molecular weight heparin for deep vein thrombosis prevention.

**Heparin**
*HEP-uh-rin*

**Heparin** and "bleedin'" sort of rhyme to remember its primary adverse effect. A student mentioned the actor Dennis Quaid's twins, who received a double dose of **heparin** that caused bleeding. Sometimes knowing a celebrity with a condition helps memory.

**Warfarin (Coumadin)**
*WAR-fa-rin (KOO-ma-din)*

That the "-parin" stem from the anticoagulants **heparin** and **enoxaparin** and "-farin" stem of **warfarin** are similar. This reminds students they are all anticoagulants. Students associate bleeding with warfare. The INR (international normalized ratio), a way of measuring **warfarin's** effectiveness, monitors the patient who is on therapy. "I-N-R" happen to be the last three letters of **warfarin**. A student said that **warfarin** has "far" in it, as in you have to go far to have blood drawn. A way to remember Vitamin K affects **Coumadin** and coagulation is to spell **Coumadin** with a "K" instead of a "C."

**Dabigatran (Pradaxa)**
*da-bih-GA-tran (pra-DAX-uh)*

Memorize **dabigatran's** "–gatran" stem to note the difference between anticoagulants. **Dabigatran** doesn't need INR monitoring like **warfarin** does. Note the last three letters in **dabigatran** as not being "I-N-R."

### ANTIPLATELET

**Clopidogrel (Plavix)**
*klo-PID-oh-grel (PLA-vix)*

**Clopidogrel** and **aspirin** work similarly leading to a reduced likelihood that platelets will stick together and clot. **Plavix** vexes platelets and keeps the blood thin.

## IX. CARDIAC GLYCOSIDE AND ANTICHOLINERGIC

A cardiac glycoside, such as **digoxin (Lanoxin)**, increases the force of contraction of the heart. We call this a positive inotropic effect. Also an antidysrhythmic, **digoxin** changes the electrochemistry of the heart to prevent dysrhythmias.

**Atropine** (**AtroPen**), an anticholinergic, prevents bradycardia, a drop in heart rate. **Atropine** can treat certain cholinergic poisonings.

## CARDIAC GLYCOSIDE

**Digoxin (Lanoxin)**
*di-JOCKS-in (la-KNOCKS-in)*

**Digoxin** treats congestive heart failure by increasing the force of the heart's contractions. **Digoxin** is derived from the plant *Digitalis lanata*. In Latin, *Digitalis* means something like hand or "digits," while *lanata* means "wooly" because the actual plant is fuzzy. Therefore, **digoxin** comes from the name *digitalis*, and brand name **Lanoxin** comes from *lanata*. Alternatively, you could remember that **Lanoxin** and **digoxin** keep your heartbeat rockin'.

## ANTICHOLINERGIC

**Atropine (AtroPen)**
ah-trow-PEEN (ah-trow-PEN)

**Atropine** causes anticholinergic (anti = against, cholinergic = of acetylcholine) effects. Anticholinergic effects fall under the broad category of "dry." Use the ABDUCT mnemonic, as in anticholinergics "abduct" water: Anhidrosis, Blurry vision (secondary to dry eyes), Dry mouth, Urinary retention, Constipation, and Tachycardia. This tachycardic side effect therapeutically prevents bradycardia in patients undergoing certain procedures.

Note: Cholinergic effects would include "wet" effects: sweating, lacrimation (watery eyes), hypersalivation, urinary incontinence, diarrhea, and bradycardia.

## CARDIO DRUG QUIZ (LEVEL 1)

Classify these drugs by placing the corresponding drug class letter next to each medication. Try to underline the stems before you start and think about the brand name and function for each drug.

1. Atorvastatin (Lipitor)
2. Clopidogrel (Plavix)
3. Enalapril (Vasotec)
4. Enoxaparin (Lovenox)
5. Furosemide (Lasix)
6. Hydrochlorothiazide (Microzide)
7. Losartan (Cozaar)
8. Metoprolol (Lopressor)
9. Nifedipine (Procardia)
10. Spironolactone (Aldactone)

**Cardio drug classes:**

A. ACE inhibitor (ACEI)
B. Alpha blocker
C. Angiotensin receptor blocker (ARB)
D. Anticoagulant
E. Antiplatelet
F. Beta blocker: selective
G. Beta blocker: non-selective
H. CCB - dihydropyridine
I. CCB - non-dihydropyridine
J. Cardiac glycoside
K. Diuretic: Loop
L. Diuretic: Osmotic
M. Diuretic: Potassium sparing
N. Diuretic: Thiazide
O. HMG-CoA reductase inhibitor
P. Vasodilator

## CARDIO DRUG QUIZ (LEVEL 2)

Classify these drugs by placing the corresponding drug class letter next to each medication. Try to underline the stems before you start and remember the brand name and function of each drug.

1. Diltiazem
2. Carvedilol
3. Olmesartan
4. Hydrochlorothiazide
5. Doxazosin
6. Amlodipine
7. Nitroglycerin
8. Lisinopril
9. Digoxin
10. Warfarin

**Cardio drug classes:**

A. ACE inhibitor (ACEI)
B. Alpha blocker
C. Angiotensin receptor blocker (ARB)
D. Anticoagulant
E. Antiplatelet
F. Beta blocker: selective
G. Beta blocker: non-selective
H. CCB - dihydropyridine
I. CCB - non-dihydropyridine
J. Cardiac glycoside
K. Diuretic: Loop
L. Diuretic: Osmotic
M. Diuretic: Potassium sparing
N. Diuretic: Thiazide
O. HMG-CoA reductase inhibitor
P. Vasodilator

# Cardio: Memorizing the Chapter

| | | |
|---|---|---|
| **Omega-3-Acid E.E.** | Atenolol | Nitroglycerin |
| **Niacin** | Metoprolol Succinate | Atorvastatin |
| **Aspirin (Low Dose)** | Metoprolol Tartrate | Rosuvastatin |
| Mannitol | Carvedilol | Fenofibrate |
| Furosemide | Enalapril | Heparin |
| Hydrochlorothiazide | Lisinopril | Enoxaparin |
| HCTZ/Triamterene | Losartan | Warfarin |
| Spironolactone | Olmesartan | Dabigatran |
| Potassium Chloride | Valsartan | Clopidogrel |
| Doxazosin | Diltiazem | Digoxin |
| Clonidine | Verapamil | Atropine |
| Propranolol | Amlodipine | |
| | Nifedipine | |

## "C" Cardio

"O" from cardio to **Omega-3 acid ethyl esters** to **niacin** to **aspirin** you need to take before the niacin to prevent flushing. Flushing five diuretics in nephron order: **mannitol, furosemide, hydrochlorothiazide, HCTZ / triamterene, spironolactone** (potassium sparing), then **potassium chloride** to alphas: alpha 1 **doxazosin**, alpha 2 **clonidine**; to betas: 1st-generation beta 1 and beta 2 **propranolol**; 2nd-generation, beta 1 only, **atenolol, metoprolol tartrate** short acting to **metoprolol succinate** long acting; to 3rd-generation **carvedilol**. Dil to pril ACEIs **enalapril, lisinopril**. ARBs "l-o-v" **losartan, olmesartan** and **valsartan**; "-sartan" to CCBs non-dihydropyridines **diltiazem** and **verapamil** to the dihydropyridines **amlodipine** and **nifedipine** vasodilating only; CCBs to **nitroglycerin**, a vasodilator. Nitroglycerin brand name

**Nitrostat** to stat-ins: **atorvastatin** to **rosuvastatin**; LDL to lowered triglycerides, **fenofibrate** fibs children tell to parents, parenteral **enoxaparin** and **heparin**, "-parin" to "-farin," **warfarin** enteral anticoagulant with **dabigatran**, "d" to **digoxin** for CHF and **atropine** to prevent a bradycardic mess.

## CARDIODE TO JOY

Cardiode to Joy, sung to the tune of Beethoven's *Ode to Joy*, is a mnemonic that you can sing, hum or just say that attaches many of the common cardio drug endings and classes to their functions.

o-l-o-l-p-r-i-l-and-s-a-r-t-a-n

be-ta-block-er-ace-in-hib-i-tor-and-ARBs-suff-ix-end

as-pir-in-and-clo-pid-o-grel-both-block-plate-lets-round-a-stent

war-fa-rin-and-hep-a-rin-are-both-an-ti-co-ag-u-lants

stat-ins-low-er-chol-est-ter-ol
dig-keeps-your-heart-from-fail-in
ver-a-pa-mil-and-am-lo-di-pine
both-block-cal-cium-chan-nels.

# CHAPTER 7 ENDOCRINE / MISC.

## I. OTC INSULIN AND EMERGENCY CONTRACEPTION

Most people don't think of insulin as an over-the-counter medication, but a prescription is not required for **regular insulin** or **NPH insulin** (intermediate duration of action) for self-pay patients. Alphabetically "N" comes before "R," but the convention is to put them in order from shortest to longest acting.

Pharmacies refrigerate insulins for stability. Insulins are expensive, so drug stores keep them in the pharmacy refrigerator not only to prevent theft, but so they know that the insulin hasn't left the refrigerator and been exposed to room temperatures. An insulin vial's box has the rectangular shape of a refrigerator to help you remember.

Emergency contraception is a form of birth control used after unprotected sex to keep a patient from getting pregnant. **Levonorgestrel (Plan B)** is a relatively recent introduction. In 1999, it was prescription only, and then it became available behind-the-counter (BTC), but is now available OTC.

**Regular insulin (Humulin R)**
*REG-you-lar IN-su-lin (HUE-myou-lin ARE)*

> **Regular insulin** is short acting, but not to be confused with the shortest-acting insulins available, such as **Humalog**. Prescribers can use **regular insulin** when patients need to adjust dosages on a sliding scale. Insulin used to come from a pig

(porcine) or cow (bovine), but now matches human insulin because of molecular engineering. Therefore, the Eli Lilly brand name **Humulin** simply squishes the words human and insulin together.

**NPH insulin (Humulin N)**
*en-pee-aitch IN-su-lin(HUE-myou-lin EN)*

The "N-P-H" in **NPH insulin** stands for neutral protamine Hagedorn. The neutral protamine refers to how Hagedorn, the inventor, chemically altered the insulin. The "e-n" pronunciation of the letter "N" sounds a little like "i-n" and can help you remember it's an intermediate acting insulin.

**Levonorgestrel (Plan B One-Step)**
*LE- vo- nor-JESS-trel (plan-bee won-step)*

You can recognize **levonorgestrel** as a progestin hormone product by the "gest" stem. Take it within 72 hours after sexual intercourse. It can cause nausea, so some flat soda might help calm this down. It's now called **Plan B One-Step** because it used to take two steps or two doses to provide this contraception.

I used to work in a college town pharmacy and every Saturday and Sunday morning, I would have a ton of students coming to pick up **Plan B**. Every time, the man drove, and the woman was in the passenger seat. When I told the male student it was fifty bucks for the Plan B, he would invariably look at her, and then pay. One time, however, I overheard her say, "Oh no, you just didn't." Remember the "g-e-s-t" from that "just" that is part of the word **Levonorgestrel**.

# II. Diabetes and Insulin

Diabetes mellitus is a condition of chronic excess blood sugar. There are three types: type I, which we previously called juvenile onset diabetes; type II, which we referred to as adult-onset diabetes; and gestational diabetes, a condition where a pregnant woman becomes diabetic. Depending on the condition, there are different drugs that can help lower blood sugar. Almost all oral medications have "gl," or "glu" for "glucose" in their names. Four of these drugs include **metformin (Glucophage), sitagliptin (Januvia), glipizide (Glucotrol)** and **glyburide (DiaBeta)**. I memorized them in alphabetical order of their drug classes: biguanide (**metformin**), dipeptidyl peptidase-4 (DPP-4) (**sitagliptin**), and sulfonylurea (**glipizide** and **glyburide**).

**Insulin** for diabetes comes from the Latin word insula, which means island. The islets of Langerhans in your pancreas have cells that produce insulin (beta cells), which lowers blood sugar; and cells that produce glucagon (alpha cells), which tells the body to raise blood glucose levels.

There are four major classes of insulin used in treatment:

- *Rapid acting* starts working in 15 minutes and lasts 4 hours, e.g., **insulin lispro (Humalog)**.
- *Short acting* works in 30 minutes and lasts about 6 to 8 hours, e.g., **regular insulin (Humulin R)**.
- *Intermediate duration* **NPH insulin (Humulin N)** works in an hour or two and lasts 14-24 hours.
- *Long duration* **insulin glargine (Lantus, Toujeo)** starts working in an hour and lasts 24 hours.

## Oral anti-diabetics - Biguanides

**Metformin (Glucophage)**
*met-FOUR-men (GLUE-co-fage)*

One student came up with this mnemonic, "If you met four men on **Glucophage**, they are diabetic then." Phagocytosis is the process of cell eating. You can use the brand name **Glucophage** to think of the medication as eating, or "phage-ing" glucose.

## Oral anti-diabetics – DPP-4 inhibitors

**Sitagliptin (Januvia)**
*sit-uh-GLIP-tin (ja-NEW-vee-uh)*

Although the "-gliptin" stem helps us recognize **sitagliptin** as an anti-diabetic, students associate the sugar you might put in "Lipton" iced tea with **sitagliptin**. The brand name **Januvia** ends in "v-i-a" and is similar to "d-i-a" from diabetes.

## Oral anti-diabetics – Sulfonylureas 2nd-generation

**Glipizide (Glucotrol)**
*GLIP-eh-zide (GLUE-co-trawl)*

The "gli-" stem in **glipizide** indicates an antihyperglycemic medication. The brand name **Glucotrol** alludes to the control of blood glucose in diabetics.

**Glyburide (DiaBeta)**
*GLY-byour-ide (die-uh-BAY-ta)*

The "gly-" stem in **glyburide** is archaic, and has been replaced in new medicines by the stem "gli-." The brand name **DiaBeta** combines the "d-i-a" from diabetic, and the "B-e-t-a" from the Beta cells, which release insulin.

## HYPOGLYCEMIA

**Glucagon (GlucaGen)**
*GLUE-ca-gone (glue-ca-JEN)*

Remember the generic with, "I use **glucagon** when the glucose is gone." **GlucaGen,** the brand name, generates glucose when a patient is hypoglycemic.

## RX INSULIN

**Insulin lispro (Humalog)**
*IN-su-lin LICE-pro (HUE-mah-log)*

**Insulin glargine** would precede **insulin lispro** alphabetically, but the convention is to list the medications shorter acting to longer acting. I use my Spanish language to help memorize that **insulin lispro** is rapid acting. When I was in Mexico on a zip line, the person in the first tower would say, "Listo, listo," meaning "You ready, I'm ready." Then I would fly fast down that zip cord. **Humalog** is a human insulin analog. I always pictured a log floating down rapids fast to remember **Humalog** is a rapid acting insulin.

**Insulin glargine (Lantus, Toujeo)**
*IN-su-lin GLAR-Jean (LAN-tuss, TWO-jzeh-oh)*

> With **insulin glargine**, I think of glaring and lurking, someone who is *slowly* creeping around. One student came up with "**Lantus** lasts all day long, take it at night, and your life will be prolonged." I've also heard "lazy **Lantus**" used to help remember it's a 24-hour drug.

## III. Thyroid hormones

Thyroid hormone stimulates the heart, metabolism, and helps with growth. A hyperthyroid patient's body uses energy too quickly because of the extra thyroid hormone in circulation.

This patient can use **propylthiouracil (PTU)** to reduce the effects of the thyroid hormone. Hypothyroid patients need extra thyroid hormone, such as **levothyroxine (Synthroid)**, for replacement.

### Hypothyroidism

**Levothyroxine (Synthroid)**
*Lee-vo-thigh-ROCKS-een (SIN-throyd)*

> The generic name **levothyroxine** has the "thyro" from thyroid in the name; you just have to remember it's for supplementation. The brand name **Synthroid** combines the words synthetic and thyroid.

### HYPERTHYROIDISM

**Propylthiouracil (PTU)**
*pro-pill-thigh-oh-YOUR-uh-sill (pee-tee-you)*

> **PTU** takes "p" from propyl, "t" from "thio," and "u" from uracil in the generic **propylthiouracil**. Although "thio" means there is a sulfur atom in the molecule, you can think of it as thyroid lowering.

## IV. HORMONES AND CONTRACEPTION

**Testosterone** is an androgen steroid hormone that naturally comes from the male testes. As a medication, prescribers use it to supplement conditions of low testosterone.

Pharmaceutical birth control, commonly known as "the pill," traditionally came from a combined oral contraceptive pill (COCP) that has a combination of an estrogen and a progestin. There are many variations of "the pill" including **Loestrin 24 Fe.** The Fe stands for ferrous, or iron, on the periodic table. The tri-phasic birth controls, such as **Tri-Sprintec**, have three different doses of an estrogen and progestin, taken variously throughout the month to mimic the body's naturally changing hormone levels. Two novel birth control delivery methods include a vaginally inserted ring **(NuvaRing)** and a transdermal patch **(OrthoEvra).**

### TESTOSTERONE

**Testosterone (AndroGel)**
*Tess-TOSS-ter-own (ANN-droh GEL)*

> Most people know the steroid hormone **testosterone**, but note the stem for a steroid is "ster."

"Andro" is the Greek prefix for male and gel is the vehicle in **AndroGel**, indicating it's a "gel" for a "male."

## Contraception – Combined oral contraceptives

**Norethindrone / ethinyl estradiol / ferrous fumarate (Loestrin 24 Fe)**
*Nor-eth-IN-drone / ETH-in-ill es-tra-DYE-all (Low-ES-trin EF-ee Twen-TEE fore)*

It's important to first memorize the estrogen stem "estr-" and progestin stem "-gest-." Then I got a little rhyme crazy with the contraceptive brand name mnemonics. One rhyme goes: "**Loestrin 24 Fe** reduces the length of menstruation, with iron supplementation, to prevent an anemic situation."

**Norgestimate / ethinyl estradiol (Tri-Sprintec)**
*Nor-JESS-teh-mate / ETH-in-ill es-tra-DYE-all"*

For most hormone-based contraceptives, it's important to note both the estrogen stem "estr-"and progestin stem "-gest-." And here's another rhyme: "**Tri-Sprintec** is triphasic, take three different doses, in seven-day spaces."

## Contraception – Patch

**Norelgestromin / ethinyl estradiol (OrthoEvra)**
*Nor-el-JESS-tro-min / ETH-in-ill es-tra-DYE-all (OR-thoe EV-rah)*

One student thought of the "Norel" in **norelgestromin** as "not oral" to remember this is a patch. Another rhyme highlights the places a woman should place the patch and length of time to leave it there: "**OrthoEvra** is a patch, put it on your arm, your abs, your buttock or back, and then take it off a week after that."

### CONTRACEPTION – RING

**Etonogestrel / ethinyl estradiol (NuvaRing)**
*Et-oh-no-JESS-trel / ETH-in-ill es-tra-DYE-all"*
*(NEW-va-ring)*

A student thought of the "Etono" in **etonogestrel** as "Eat, oh no" to remember it's not an oral tablet. A final rhyme and then I promise, no more: "**NuvaRing** is one way to keep the stork away before you are ready for parenting a bay-bay" (By the way, just as an aside, there is no time when you are actually ready for parenting.)

## V. OVERACTIVE BLADDER, URINARY RETENTION, ERECTILE DYSFUNCTION, BENIGN PROSTATIC HYPERPLASIA

Frequently the words incontinence, urinary retention, impotence, and benign prostatic hyperplasia are confused:

- **Overactive bladder (OAB) (incontinence)** is an inability to retain urine due to overactive bladder.
- **Urinary retention** is a difficulty in urination.
- **Erectile dysfunction (ED) (impotence)** is the inability to achieve or maintain an erection.
- **Benign prostatic hyperplasia (BPH)** is a benign (not harmful) prostate growth or increase in size.

We try not to use the terms "incontinence" or "impotence" because of the harshness to the words.

## OVERACTIVE BLADDER

**Oxybutynin (Ditropan, Oxytrol OTC)**
*ox-e-BYOU-tin-in (DIH-trow-pan)*

> One student remembered **oxybutynin** as keepin' the urin' in. Both **Ditropan** and **Oxytrol** have the "t-r-o" from control for controlling an overactive bladder.

**Solifenacin (VESIcare)**
*sol-eh-FEN-a-sin (VEH-si-care)*

> **Solifenacin** is given once daily, so thinking about it as "slow-fenacin" helps in remembering this point. Also, **solifenacin** solves the problem of urine that needs to be "fenced in." **VESIcare** contains *vesica,* which means "bladder" in Latin.

**Tolterodine (Detrol)**
*toll-TER-oh-dean (DEH-trawl)*

> The generic name **tolterodine** has the "t-r-o" from control in the name as well. **Detrol** helps control the detrusor muscle, keeping urine in.

## URINARY RETENTION

**Bethanechol (Urecholine)**
*beh-THAN-uh-call (yur-eh-CO-lean)*

> The "chol" in **bethanechol** helps you remember it's cholinergic. While anticholinergics are dry, cholinergics do the opposite and make things wet.

This drug assists the bladder muscles in expelling urine. The brand name **Urecholine** alludes to how it affects urination through cholinergic effects.

## Erectile dysfunction (PDE-5 inhibitors)

**Sildenafil (Viagra)**
*sill-DEN-uh-fill (vie-AG-rah)*

There is a scene with Jack Nicholson in the movie *Something's Gotta Give* that reminds us **sildenafil** shouldn't be used with nitrates like **nitroglycerin**. **Viagra** brings viable growth - an erection.

**Tadalafil (Cialis)**
*ta-DAL-uh-fill (see-AL-is)*

**Cialis** is the weekend pill because it, unlike **sildenafil**, lasts the weekend, as it has a long half-life. I asked some students about how they remembered **tadalafil** and I'm hesitant to share their mnemonic. One said you just think "ta-dah" as in "surprise." I stopped them before they started on to how they remember the "fil" part of the generic name. **Cialis** is the dual bathtub commercials drug.

## BPH – Alpha blocker

**Tamsulosin (Flomax)**
*tam-syoo-LOW-sin (FLOW-Max)*

**Flomax** allows for maximum urinary flow. The "osin" ending is not an actual stem, but a way to connect **tamsulosin** and **alfuzosin** as being similar.

**Alfuzosin (Uroxatral)**
*al-fyoo-ZOH-sin (YUR-ox-uh-trall)*

With BPH, there is sometimes difficulty with urine control and I think the brand **Uroxatral** sounds a little like "Urine control."

## BPH – 5-ALPHA-REDUCTASE INHIBITOR

**Dutasteride (Avodart)**
*due-TAS-ter-ide (AH-vo-dart)*

Use the "–steride" stem to recognize the 5-alpha-reductase inhibitors like **dutasteride**. The "ster" for steroid helps you remember it's for men (and prostate).

**Finasteride (Proscar, Propecia)**
*fin-AS-ter-ide" (PRO-scar, pro-PEE-shuh)*

To connect the brand name **Proscar** to the generic **finasteride**, a professor told me of a student who used the phrase "That pro's car is the finest ride." **Proscar** is for prostate **care**. **Finasteride's** other brand name, **Propecia**, alludes to hair growth and is to reverse of alopecia (hair loss).

## ENDOCRINE / MISC. DRUG QUIZ (LEVEL 1)

Classify these drugs by placing the corresponding drug class letter next to each medication. Try to underline the stems before you start and think about the brand name and function of each drug.

1. Glipizide (Glucotrol)
2. Glucagon (GlucaGen)
3. Glyburide (DiaBeta)
4. Insulin glargine (Lantus)
5. Levothyroxine (Synthroid)
6. Metformin (Glucophage)
7. Propylthiouracil (PTU)
8. Regular insulin (Humulin R)
9. Solifenacin (VESIcare)
10. Sildenafil (Viagra)

**Endocrine system drug classes:**

A. Anti-diabetic
B. BPH - 5-alpha-reductase inhibitor
C. BPH - alpha blocker
D. Contraception – COCP triphasic
E. Contraception – COCP with iron
F. Contraception – patch
G. Contraception – ring
H. For hypoglycemia
I. For hypothyroidism
J. For hyperthyroidism
K. Erectile dysfunction
L. OAB
M. Insulin – long acting
N. Insulin – short acting
O. Urinary retention

## ENDOCRINE / MISC. DRUG QUIZ (LEVEL 2)

Classify these drugs by placing the corresponding drug class letter next to each medication. Try to underline the stems before you start and remember the brand name and function of each drug.

1. Dutasteride
2. Tamsulosin
3. Tolterodine
4. Tadalafil
5. Norethindrone / ethinyl estradiol / Fe
6. Norelgestromin / ethinyl estradiol
7. Oxybutynin
8. Bethanechol
9. Sildenafil
10. Finasteride

**Endocrine system drug classes:**

A. Anti-diabetic
B. BPH - 5-alpha-reductase inhibitor
C. BPH - alpha blocker
D. Contraception – COCP triphasic
E. Contraception – COCP with iron
F. Contraception – patch
G. Contraception – ring
H. For hypoglycemia
I. For hypothyroidism
J. For hyperthyroidism
K. Erectile dysfunction
L. OAB
M. Insulin – long acting
N. Insulin – short acting
O. Urinary retention

# ENDOCRINE: MEMORIZING THE CHAPTER

**Regular Insulin**
**NPH Insulin**
**Levonorgestrel**
Metformin
Sitagliptin
Glipizide
Glyburide
Glucagon
Insulin lispro
Insulin glargine
Levothyroxine
Propylthiouracil
Testosterone

Ethinyl Estradiol / Norethindrone / Ferrous fumarate (Loestrin 24 Fe)
Ethinyl Estradiol / Norgestimate (Tri-Sprintec)
Ethinyl Estradiol / Etonogestrel (NuvaRing)
Ethinyl Estradiol / Norelgestromin (OrthoEvra)

Oxybutynin
Solifenacin
Tolterodine
Bethanechol
Sildenafil
Tadalafil
Alfuzosin
Tamsulosin
Dutasteride
Finasteride

(Oral contraceptives brand names in parenthesis for clarity)

## "E" ENDOCRINE

Start with the OTC two middle peak insulins from the "r" and "n" in endocrine: **regular insulin** and **insulin NPH**; then Plan B One-Step **levonorgestrel**. RX: four oral antidiabetics in order of drug class: the biguanide **metformin,** the DPP-4 **sitagliptin,** and two sulfonylureas **glipizide** and **glyburide**; then **glucagon** for when the glucose is gone. On to prescription shortest and longest acting insulins: **insulin aspart (Humalog)** and **insulin glargine (Lantus).** "T" in lantus to "T" thyroid level low (**levothyroxine**), to thyroid level high **(PTU).** "T" again to low T **testosterone**. Testosterone "-ster-" stem to estrogen "estr-" stem. **Ethinyl estradiol** "estr-" times four in four oral contraceptives, from high on the body to low starting with an oral (po) form with iron, another p.o. tri-phasic, a belly patch, and vaginal ring, brand names are easier first, Loestrin 24 Fe, Tri-Sprintec, OrthoEvra patch, NuvaRing. Generic progestins are next: **norethindrone**, **norgestimate**, **norelgestromin**, **etonogestrel**. Overactive bladder (OAB) **oxybutynin, solifenacin**, and **tolterodine**, to the opposite – urinary retention: **bethanechol**; to a lack of erection by half-life: **sildenafil** short and **tadalafil** long, but also **tadalafil** for BPH and alpha-blockers **alfuzosin** and **tamsulosin**, and 5-alpha reductase **dutasteride** and **finasteride**.

# CHAPTER 8 MATCHING EXAMS

Congratulations! You've made it through the first 200 drugs in this book. I hope that you've started to commit the mnemonics to memory and found suffixes and prefixes you can count on to help you remember drug classes.

The following pages include two final exams. The first final exam provides generic and brand names for the medications, and is a little easier. The second final exam, like some of your board exams, provides only generic names for a greater challenge.

If you feel you want more of a review, you can look at the summary of prefixes and suffixes in the index and then move on to the exams.

Remember to underline the drug stems first and picture what chapter you learned them in.

After you are done, review the ones you missed and shore up some memorization gaps. Then you are ready to go from 200 drugs to 350 drugs (if you really want to).

## MATCHING EXAM 1, QUESTIONS 1-25, CH. 1-3

___1 Loperamide (Imodium)
___2 Acetaminophen (Tylenol)
___3 Allopurinol (Zyloprim)
___4 APAP / Codeine (Tylenol/Codeine)
___5 Budesonide / Formoterol (Symbicort)
___6 Omeprazole (Prilosec)
___7 Esomeprazole (Nexium)
___8 Febuxostat (Uloric)
___9 Guaifenesin / Dextromethorphan (Robitussin DM)
__10 Fluticasone / Salmeterol (Advair)
__11 Hydrocodone / APAP (Vicodin)
__12 Fentanyl (Duragesic)
__13 Famotidine (Pepcid)
__14 Docusate sodium (Colace)
__15 Celecoxib (Celebrex)
__16 Aspirin [ASA] (Ecotrin)
__17 Alendronate (Fosamax)
__18 Albuterol (ProAir)
__19 ASA / APAP / Caffeine (Excedrin)
__20 Bismuth subsalicylate (Pepto-Bismol)
__21 Calcium carbonate (Tums)
__22 Guaifenesin / Codeine (Cheratussin AC)
__23 Cetirizine (Zyrtec)
__24 Diphenhydramine (Benadryl)
__25 Etanercept (Enbrel)

a. 1st-generation antihistamine
b. 2nd-generation antihistamine
c. $5\text{-}HT_3$ receptor antagonist
d. Antacid
e. Anticholinergic for asthma
f. Anti-diarrheal
g. Anti-gout
h. Anti-nausea
i. Bisphosphonate
j. DMARD
k. $H_2$ blocker
l. Laxative
m. Mucolytic/cough
n. Non-narcotic analgesic combo
o. Non-narcotic analgesic
p. NSAID
q. Opioid analgesic
r. Proton pump inhibitor
s. Short-acting bronchodilator
t. Steroid/bronchodilator

## MATCHING EXAM 1, QUESTIONS 26-50, CH. 4

__26 Raltegravir (Isentress)
__27 Amoxicillin / Clavulanate (Augmentin)
__28 Ceftriaxone (Rocephin)
__29 Doxycycline (Doryx)
__30 Levofloxacin (Levaquin)
__31 Sulfamethoxazole / Trimethoprim (Bactrim)
__32 Cephalexin (Keflex)
__33 Ethambutol (Myambutol)
__34 Rifampin (Rifadin)
__35 Isoniazid (INH)
__36 Amikacin (Amikin)
__37 Cefepime (Maxipime)
__38 Amphotericin B (Fungizone)
__39 Amoxicillin (Amoxil)
__40 Azithromycin (Zithromax)
__41 Fluconazole (Diflucan)
__42 Oseltamivir (Tamiflu)
__43 Minocycline (Minocin)
__44 Clarithromycin (Biaxin)
__45 Gentamicin (Garamycin)
__46 Pyrazinamide (PZA)
__47 Nystatin (Mycostatin)
__48 Erythromycin (E-Mycin)
__49 Ciprofloxacin (Cipro)
__50 Acyclovir (Zovirax)

a. 1$^{st}$-generation cephalosporin
b. 2$^{nd}$-generation cephalosporin
c. 3$^{rd}$-generation cephalosporin
d. 4$^{th}$-generation cephalosporin
e. Antibiotic: aminoglycoside
f. Antibiotic: fluoroquinolone
g. Antibiotic: macrolide
h. Antibiotic: penicillin
i. Antibiotic: sulfa
j. Antibiotic: tetracycline
k. Antifungal
l. Antituberculosis
m. Antiviral: herpes
n. Antiviral: HIV
o. Antiviral: Influenza

## MATCHING EXAM 1, QUESTIONS 51-75, CH. 5

__51 Sertraline (Zoloft)
__52 Carbamazepine (Tegretol)
__53 Cyclobenzaprine (Flexeril)
__54 Divalproex (Depakote)
__55 Gabapentin (Neurontin)
__56 Lithium (Lithobid)
__57 Meclizine (Antivert)
__58 Lidocaine (Solarcaine)
__59 Fluoxetine (Prozac)
__60 Escitalopram (Lexapro)
__61 Clonazepam (Klonopin)
__62 Benzocaine (Anbesol)
__63 Amitriptyline (Elavil)
__64 Atomoxetine (Strattera)
__65 Dexmethylphenidate (Focalin)
__66 Escitalopram (Lexapro)
__67 Haloperidol (Haldol)
__68 Lorazepam (Ativan)
__69 Levodopa / Carbidopa (Sinemet)
__70 Donepezil (Aricept)
__71 Citalopram (Celexa)
__72 Alprazolam (Xanax)
__73 Chlorpromazine (Thorazine)
__74 Eszopiclone (Lunesta)
__75 Isocarboxazid (Marplan)

a. ADHD drug/non-stimulant
b. ADHD drug/stimulant
c. Alzheimer's
d. Antidepressant: MAOI
e. Antidepressant: SNRI
f. Antidepressant: SSRI
g. Antidepressant: TCA
h. Antiepileptic: newer
i. Antiepileptic: traditional
j. Antipsychotic: $2^{nd}$ generation
k. Antipsychotic: $1^{st}$-generation
l. Benzodiazepine
m. Local anesthetic
n. Muscle relaxer
o. Parkinson's
p. Sedative-hypnotic
q. Simple salt
r. Vertigo/motion sickness

## MATCHING EXAM 1, QUESTIONS 76-100, CH. 6-7

__76 Norethindrone / Ethinyl estradiol / Fe (Loestrin Fe)
__77 Bethanechol (Urecholine)
__78 Amlodipine (Norvasc)
__79 Diltiazem (Cardizem)
__80 Glucagon (GlucaGen)
__81 Finasteride (Proscar)
__82 Hydrochlorothiazide (Microzide)
__83 Regular insulin (Humulin R)
__84 HCTZ / Triamterene (Dyazide)
__85 Oxybutynin (Ditropan)
__86 Clopidogrel (Plavix)
__87 Norgestimate / Ethinyl estradiol (Tri-Sprintec)
__88 Propranolol (Inderal)
__89 Furosemide (Lasix)
__90 Enalapril (Vasotec)
__91 Insulin glargine (Lantus)
__92 Enoxaparin (Lovenox)
__93 Atorvastatin (Lipitor)
__94 Digoxin (Lanoxin)
__95 Norelgestromin / Ethinyl estradiol (OrthoEvra)
__96 Dutasteride (Avodart)
__97 Glipizide (Glucotrol)
__98 Heparin
__99 Glyburide (DiaBeta)
_100 Etonogestrel / Ethinyl estradiol (NuvaRing)

a. ACE inhibitor
b. ARB
c. Anticoagulant
d. Antidiabetic
e. Antiplatelet
f. Beta blocker
g. BPH: 5-alpha-reductase inhibitor
h. BPH: Alpha blocker
i. Calcium channel blocker
j. Cardiac glycoside
k. Contraception
l. Diuretic
m. For hypoglycemia
n. For hyperthyroidism
o. For hypothyroidism
p. HMG-CoA reductase inhib.
q. Erectile dysfunction
r. OAB
s. Longer duration insulin
t. Slower acting insulin
u. Urinary retention
v. Vasodilator

## MATCHING EXAM 2, QUESTIONS 1-25, CH. 1-3

___1 Loratadine
___2 Diphenhydramine
___3 Ranitidine
___4 Pseudoephedrine
___5 Esomeprazole
___6 Naproxen
___7 Triamcinolone
___8 Infliximab
___9 Loperamide
__10 Bismuth subsalicylate
__11 Ibuprofen
__12 Magnesium hydroxide
__13 Methotrexate [MTX]
__14 Cetirizine
__15 Hydrocodone / APAP
__16 Prednisone
__17 Sumatriptan
__18 Tiotropium
__19 Promethazine
__20 Polyethylene glycol
__21 Oxycodone / APAP
__22 Omeprazole
__23 Ondansetron
__24 Morphine
__25 Methylprednisolone

a. 1st-generation antihistamine
b. 2nd-generation antihist.
c. 5-$HT_3$ receptor antagonist
d. Antacid
e. Anticholinergic for asthma
f. Anti-diarrheal
g. Anti-gout
h. Anti-nausea
i. Decongestant
j. DMARD
k. $H_2$ blocker
l. Laxative
m. Mucolytic/cough
n. Non-narcotic analg. combo
o. Non-narcotic analgesic
p. NSAID
q. Opioid analgesic
r. Proton pump inhibitor
s. Steroid
t. Ulcerative colitis

## MATCHING EXAM 2, QUESTIONS 26-50, CH. 4

__26 Doxycycline
__27 Amoxicillin
__28 SMZ / TMP
__29 Amphotericin B
__30 Erythromycin
__31 Ceftriaxone
__32 Rifampin
__33 Isoniazid
__34 Cephalexin
__35 Ethambutol
__36 Nystatin
__37 Minocycline
__38 Valacyclovir
__39 Zanamivir
__40 Amikacin
__41 Gentamicin
__42 Clarithromycin
__43 Cefepime
__44 Oseltamivir
__45 Fluconazole
__46 Ciprofloxacin
__47 Amoxicillin / Clavulanate
__48 Levofloxacin
__49 Pyrazinamide
__50 Darunavir

a. $1^{st}$-generation cephalosporin
b. $2^{nd}$-generation cephalospor.
c. $3^{rd}$-generation cephalosporin
d. $4^{th}$-generation cephalospor.
e. Antibiotic: aminoglycoside
f. Antibiotic: fluoroquinolone
g. Antibiotic: macrolide
h. Antibiotic: penicillin
i. Antibiotic: sulfa
j. Antibiotic: tetracycline
k. Antifungal
l. Antituberculosis
m. Antiviral: herpes
n. Antiviral: HIV
o. Antiviral: influenza

## MATCHING EXAM 2, QUESTIONS 51-75, CH. 5

__51 Selegiline
__52 Sertraline
__53 Quetiapine
__54 Escitalopram
__55 Meclizine
__56 Isocarboxazid
__57 Gabapentin
__58 Levodopa / Carbidopa
__59 Memantine
__60 Phenytoin
__61 Ramelteon
__62 Citalopram
__63 Venlafaxine
__64 Trazodone
__65 Pregabalin
__66 Fluoxetine
__67 Lorazepam
__68 Haloperidol
__69 Lidocaine
__70 Paroxetine
__71 Risperidone
__72 Zolpidem
__73 Scopolamine
__74 Methylphenidate
__75 Lithium

a. ADHD drug / non-stimulant
b. ADHD drug/stimulant
c. Alzheimer's
d. Antidepressant: MAOI
e. Antidepressant: SNRI
f. Antidepressant: SSRI
g. Antidepressant: TCA
h. Antiepileptic: Newer
i. Antiepileptic: Traditional
j. Antipsychotic: $2^{nd}$-generat.
k. Antipsychotic: $1^{st}$-generation
l. Benzodiazepine
m. Local anesthetic
n. Muscle relaxer
o. Parkinson's
p. Sedative-hypnotic
q. Simple salt
r. Vertigo / motion sickness

## MATCHING EXAM 2, QUESTIONS 76-100, CH. 6-7

__76 Levothyroxine
__77 Spironolactone
__78 Tolterodine
__79 Rosuvastatin
__80 Carvedilol
__81 Olmesartan
__82 Lovastatin
__83 Lisinopril
__84 Metformin
__85 Oxybutynin
__86 Atorvastatin
__87 Valsartan
__88 Tamsulosin
__89 Nifedipine
__90 Propylthiouracil
__91 Losartan
__92 Propranolol
__93 Tadalafil
__94 Sildenafil
__95 Verapamil
__96 Glipizide
__97 Enalapril
__98 Mannitol
__99 Metoprolol
__100. Warfarin

a. ACE inhibitor
b. Angiotensin receptor blocker
c. Anticoagulant
d. Antidiabetic
e. Antiplatelet
f. Beta blocker
g. BPH: 5-alpha-reductase inhib.
h. BPH: Alpha blocker
i. Calcium channel blocker
j. Cardiac glycoside
k. Contraception
l. Diuretic
m. For hypoglycemia
n. For hyperthyroidism
o. For hypothyroidism
p. HMG-CoA reductase inhibitor
q. Erectile dysfunction
r. OAB
s. Longer duration insulin
t. Slower acting insulin

# CHAPTER 9 LEARNING 350 DRUGS

## BUILDING ON WHAT YOU'VE LEARNED

Now that you have the list of 200 memorized, you can supplement it. I'm going to add 150 more drugs so you can see how the first 200 can act as a lattice on which to build. I've left the original 200 in un-bolded type and **bolded** the new ones, with a discussion following each list regarding my own rationale.

## CHAPTER 1 – GASTROINTESTINAL

### MEDICATIONS

**I. Peptic Ulcer Disease**

Antacids
1. Calcium Carbonate (Tums)
2. Magnesium Hydroxide (Milk of Magnesia)

Histamine-2 Receptor Antagonists ($H_2$RAs)
3. Famotidine (Pepcid)
4. Ranitidine (Zantac)

Proton Pump Inhibitors (PPIs)
**5. Dexlansoprazole (Dexilant)**
6. Esomeprazole (Nexium)
7. Omeprazole (Prilosec)
**8. Lansoprazole (Prevacid)**
**9. Pantoprazole (Protonix)**
**10. Rabeprazole (AcipHex)**

**II. Diarrhea, constipation, and emesis**

Antidiarrheals
11. Bismuth Subsalicylate (Pepto-Bismol)
12. Loperamide (Imodium)
**13. Diphenoxylate / atropine (Lomotil)**

Constipation - Stool softener
14. Docusate sodium (Colace)

Constipation - Osmotic
15. Polyethylene glycol (PEG) 3350 (MiraLax)

Constipation - Miscellaneous
**16. Lubiprostone (Amitiza)**

Antiemetic - Serotonin 5-$HT_3$ receptor antagonist
17. Ondansetron (Zofran)

Antiemetic - Phenothiazine
18. **Prochlorperazine (Compazine)**
19. Promethazine (Phenergan)

**III. Autoimmune disorders**

Ulcerative colitis
20. Infliximab (Remicade)

## Discussion

With the proton pump inhibitors, **esomeprazole** (the (S) enantiomer of **omeprazole**) and **dexlansoprazole** (the (R)(+) enantiomer of **lansoprazole**) could be considered superior and segregated. Nevertheless, I just to put the PPIs in alphabetical order by generic name: **dexlansoprazole**

**(Dexilant)**, **esomeprazole (Nexium)**, **omeprazole (Prilosec)**, **lansoprazole (Prevacid)**, **pantoprazole (Protonix)**, and **rabeprazole (AcipHex)**. Some of these newer brand names cleverly hint at function: **Prevacid** will "prevent acid," **Protonix** "nixes protons," **AcipHex** combines "a-c-i" from acid, "pH" from the pH scale, and "ex" meaning to get rid of.

Under antidiarrheals, **diphenoxylate with atropine (Lomotil)** would alphabetically precede **loperamide**. I put it *after* **loperamide** because **diphenoxylate with atropine** represents an increase in the aggressiveness of treatment from OTC to prescription. The brand name **Lomotil** spells out "low motility" for slowing down diarrhea. The **atropine** is there to prevent someone from crushing the **diphenoxylate** and injecting it illicitly.

I would also put **lubiprostone (Amitiza)** after **docusate sodium** and **polyethylene glycol** because it's a prescription item and would represent an escalation in the aggressiveness of treatment for constipation. While the "-prost-" stem indicates prostaglandin, it doesn't really help with immediate therapeutic recognition. With the laxative **lubiprostone**, I would think of using lube to propel a stone out of the body. **Prochlorperazine (Compazine)** and **promethazine (Phenergan)** share the same first three letters, "p-r-o," and last five letters, "a-z-i-n-e." While this isn't a stem, both are phenothiazines. By memorizing these two drugs in alphabetical order, you can use this similarity to recognize their comparable antiemetic function. Both happen to be available in suppository formulations as well.

# CHAPTER 2 – MUSCULOSKELETAL

## MEDICATIONS

### I. NSAIDs and pain

OTC Analgesics - NSAIDs
21. Aspirin [ASA] (Ecotrin)
22. Ibuprofen (Advil, Motrin)
23. Naproxen (Aleve)

OTC Analgesic - Non-narcotic
24. Acetaminophen [APAP] (Tylenol)

OTC Migraine - NSAID / Non-narcotic analgesic
25. ASA/APAP/Caffeine (Excedrin Migraine)

RX Migraine - Narcotic and Non-narcotic analgesic
**26. Butalbital / APAP / Caffeine (Fioricet)**

RX Analgesics - NSAIDs
**27. Diclofenac sodium extended release (Voltaren XR)**
**28. Etodolac (Lodine)**
**29. Indomethacin (Indocin)**
30. Meloxicam (Mobic)
**31. Nabumetone (Relafen)**

RX Analgesics - NSAIDs - COX-2 inhibitor
32. Celecoxib (Celebrex)

### II. Opioids and narcotics

Opioid analgesics - Schedule II
33. Morphine (Kadian, MS Contin)

34. Fentanyl (Duragesic, Sublimaze)
35. Hydrocodone / Acetaminophen (Vicodin)
**36. Hydrocodone / Chlorpheniramine (Tussionex)**
**37. Hydrocodone / Ibuprofen (Vicoprofen)**
**38. Methadone (Dolophine)**
**39. Oxycodone (OxyIR, Oxycontin)**
40. Oxycodone / Acetaminophen (Percocet)

Opioid analgesics - Schedule III
41. Acetaminophen w/codeine (Tylenol/codeine)

Mixed-opioid receptor analgesic – Schedule IV
42. Tramadol (Ultram)
**43. Tramadol / Acetaminophen (Ultracet)**

Opioid antagonist
44. Naloxone (Narcan)
**45. Buprenorphine / Naloxone (Suboxone)** [CIII]

## III. Headaches and migraine

5-$HT_1$ receptor agonist
46. Eletriptan (Relpax)
47. Sumatriptan (Imitrex)

## IV. DMARDs and rheumatoid arthritis

48. Methotrexate (Rheumatrex)
49. Abatacept (Orencia)
50. Etanercept (Enbrel)

## V. Osteoporosis

Bisphosphonates
51. Alendronate (Fosamax)
52. Ibandronate (Boniva)

**53. Risedronate (Actonel)**

**VI. Selective estrogen receptor modulator (SERM)**

**54. Raloxifene (Evista)**

**VII. Muscle relaxants**

**55. Baclofen (Lioresal)**
**56. Carisoprodol (Soma)**
57. Cyclobenzaprine (Flexeril)
58. Diazepam (Valium)
**59. Metaxalone (Skelaxin)**
**60. Methocarbamol (Robaxin)**
**61. Tizanidine (Zanaflex)**

**VIII. Gout**

**62. Colchicine (Colcrys)**

Uric acid reducers
63. Allopurinol (Zyloprim)
64. Febuxostat (Uloric)

## Discussion

Adding **butalbital**, a barbiturate, to **acetaminophen** and **caffeine** in **Fioricet** provides an escalation from the OTC **aspirin/acetaminophen/caffeine** combination in **Excedrin Migraine**. I pair those two drugs in my mind to help remember their therapeutic function.

For the NSAIDs I simply alphabetized four additions around **meloxicam. Diclofenac sodium extended-release (Voltaren XR)** is a generic name derived from its chemical structure 2-(2,6-dichloranilino) phenylacetic acid with a change from "p-h" to "f" in the middle. **Etodolac (Lodine)** shares the same generic stem, "ac." The manufacturers of **indomethacin** simply removed the middle "metha" to get the brand name **Indocin. Nabumetone's** brand name **Relafen** sounds a little like getting pain relief with an "en-said."

In combination medications, I put the shared generic drug first, followed by the additional drug. After **hydrocodone / acetaminophen (Vicodin)**, I followed with the **hydrocodone / chlorpheniramine (Tussionex)** and **hydrocodone / ibuprofen (Vicoprofen). Tussionex** is a pineapple-flavored liquid antitussive, hence the brand name. **Vicoprofen** is like **Vicodin**, but with **ibuprofen** instead of **acetaminophen**. Therefore, the manufacturer replaced the "i-b-u" of **ibuprofen** with "V-i-c-o" of **Vicodin.**

I put **methadone (Dolophine)** in alphabetically although its primary purpose is to help patients addicted to opiates get off narcotics, not provide pain relief. **Oxycodone** by itself comes in an immediate release form, **OxyIR**, and an extended release form, **OxyContin**. The "c-o-n-t-i-n" means continuous release. Many people mispronounce this as oxy-cotton, using a word they are familiar with.

**Tramadol** with **acetaminophen (Ultracet)** follows **tramadol** by itself. The manufacturer took the brand name **Ultram** (for **tramadol** alone), dropped the "m" and added "acet" from **acetaminophen**.

Patients use **buprenorphine / naloxone (Suboxone)** like **methadone** to help detox from opiate addiction. The **naloxone** is there to keep patients from crushing the drug and injecting it.

**Risedronate (Actonel)** is simply another bisphosphonate that you can recognize from the "-dronate" stem. We classify the selective estrogen receptor modulator (SERM) **raloxifene (Evista)** as an antiestrogen by its "–ifene" stem.

There are many muscle relaxants and I have simply alphabetized them by generic name: **baclofen (Lioresal)**, **carisoprodol (Soma)**, **metaxalone (Skelaxin)**, **methocarbamol (Robaxin)**, and **tizanidine (Zanaflex)**. Some of the brand names hint at muscle relaxation, e.g., **Soma** sounds like somnolence (sleepiness), and if you are a literature nerd, this drug was a hallucinogen in Aldous Huxley's 1932 book, *Brave New World.* **Skelaxin** alludes to "skeletal relaxin'," **Robaxin** and relaxin' go together, and **Zanaflex** ends with "flex" for increased flexibility.

**Colchicine (Colcrys)** relieves an acute gouty attack, so I put that before the uric acid reducers that treat chronic increased uric acid. The "crys" in **Colcrys** sounds like the painful gouty crystals that often form in the big toe. Google "gouty crystal" images and you'll see that they look like needles.

# Chapter 3 – Respiratory

## Medications

### I. Antihistamines and decongestants

Antihistamine - 1st-generation
65. Diphenhydramine (Benadryl)
**66. Hydroxyzine (Atarax)**

OTC Antihistamine - 2nd-generation
67. Cetirizine (Zyrtec)
68. Loratadine (Claritin)

OTC Antihistamine - 3nd generation
**69. Fexofenadine (Allegra)**
**70. Levocetirizine (Xyzal)**

OTC Antihistamine - Eye Drops
**71. Olopatadine (Patanol, Pataday)**

Antihistamine - Nasal Spray
**72. Azelastine (Astelin)**

OTC Antihistamine - 2nd generation / Decongestant
73. Loratadine-D (Claritin-D)

BTC/OTC Decongestants
74. Pseudoephedrine (Sudafed) [BTC]
75. Phenylephrine (NeoSynephrine) [OTC]
76. Oxymetazoline (Afrin) [OTC]

## II. Allergic rhinitis steroid, antitussives, and mucolytics

Allergic rhinitis steroid
**77. Mometasone nasal inhaler (Nasonex)**
78. Triamcinolone (Nasacort Allergy 24HR)

OTC Antitussive / Mucolytic
79. Guaifenesin/DM (Mucinex DM, Robitussin DM)

RX Antitussive / Mucolytic
80. Guaifenesin / Codeine (Cheratussin AC)

RX Antitussive
**81. Benzonatate (Tessalon Perles)**

## III. Asthma

Oral steroids
**82. Dexamethasone (Decadron)**
83. Methylprednisolone (Medrol)
84. Prednisone (Deltasone)

Ophthalmic steroid
**85. Loteprednol ophthalmic (Lotemax)**

Inhaled steroid / Beta$_2$ receptor agonist
86. Budesonide / Formoterol (Symbicort)
87. Fluticasone / Salmeterol (Advair)

Inhaled steroid
**88. Budesonide (Rhinocort, Pulmicort Flexhaler)**
89. Fluticasone (Flonase, Flovent HFA, Flovent Diskus)

Beta$_2$ receptor agonist short acting
90. Albuterol (ProAir HFA, Proventil)

**91. Levalbuterol (Xopenex HFA)**

Beta$_2$ receptor agonist / Anticholinergic
92. Albuterol / Ipratropium (DuoNeb)
**93. Albuterol / Ipratropium (Combivent)**

Anticholinergic
94. Tiotropium (Spiriva)

Leukotriene receptor antagonist
95. Montelukast (Singulair)

Anti-IgE antibody
96. Omalizumab (Xolair)

**IV. Anaphylaxis**

97. Epinephrine (EpiPen)

## DISCUSSION

I added the first-generation antihistamine **hydroxyzine (Atarax)** after **diphenhydramine (Benadryl).** The "x-y-z" inside **hydroxyzine** matches the brand name of a third-generation antihistamine **Xyzal**. The third-generation antihistamine **fexofenadine (Allegra)** is a safe active metabolite of **terfenadine (Seldane),** a drug the manufacturer removed from the market because of cardiac side effects. **Levocetirizine (Xyzal)** is not the active metabolite, but rather the left enantiomer of **cetirizine (Zyrtec).**

From oral antihistamines, I moved to ophthalmic (eye) and nasal forms. **Olopatadine (Patanol, Pataday)** shares the

"-atadine" antihistamine stem of **loratadine**. The two "o's" at the beginning of the generic **olopatadine** look like eyes. **Azelastine (Astelin)** is a medicine you stick in your nose. Prescription drugs sometimes transition to over-the-counter (OTC), so I just put **mometasone (Nasonex),** alphabetically before another allergic rhinitis steroid **triamcinolone (Nasacort Allergy 24 HR)**.

**Benzonatate (Tessalon Perles)** doesn't work like codeine, but as a local anesthetic to decrease the sensitivity of lung receptors, reducing your need to cough. The "Tess" looks like "tussive" and you can think of getting a pearl (as from a oyster) stuck in your throat. This drug is not in any way an anxiolytic (benzodiazepine), but it's interesting that the name contains the letters "b-e-n-z-o."

**Dexamethasone (Decadron)** is another oral steroid. The ophthalmic preparation **loteprednol (Lotemax)** has two o's for eyes, and comes after the two "pred" oral steroids, **methylprednisolone** and **prednisone**.

**Budesonide (Rhinocort, Pulmicort flexhaler)** bears a resemblance to **fluticasone** in that it comes as an over-the-counter nasal spray and prescription inhaler. "Rhino" is for nose and "cort" is for corticosteroid.

**Levalbuterol (Xopenex)** is the enantiomer of **albuterol**, so it should work a little bit better. I think of "open" and "exhale" when I see the brand name **Xopenex**. **DuoNeb** and **Combivent** have identical ingredients, but I wanted to make clear **DuoNeb** is a nebulization solution and **Combivent** is a combined inhaler.

# CHAPTER 4 – IMMUNE

## MEDICATIONS

### I. OTC Antimicrobials

Antibiotic cream
98. Neomycin / Polymyxin B / Bacitracin (Neosporin)
**99. Mupirocin (Bactroban) [RX]**

Antifungal cream
100. Butenafine (Lotrimin Ultra)
**101. Terbinafine (Lamisil)**
**102. Clotrimazole / Betamethasone (Lotrisone)** [RX]

Vaccinations [Some RX, some antibacterial]
**103. Diphtheria toxoid (Boostrix)**
**104. Haemophilus influenzae Type B (Pedvax HIB)**
105. Influenza vaccine (Fluzone, Flumist)
**106. Measles, Mumps and Rubella (MMR)**
**107. Meningococcal (conjugate and polysaccharide) (Menomune)**
**108. Pertussis in combination**
**109. Pneumococcal (conjugate and polysaccharide) (Prevnar 13, Pneumovax 23)**
**110. Polio**
**111. Rotavirus (RotaTeq)**
**112. Tetanus in combination**
**113. Varicella (Varivax)**
**114. Zoster (Zostavax)**

Antiviral OTC
115. Docosanol (Abreva)

## II. Antibiotics that affect the cell wall

Penicillins
116. Amoxicillin (Amoxil)
**117. Penicillin (Veetids)**

Penicillin/Beta-lactamase inhibitor
118. Amoxicillin / Clavulanate (Augmentin)

Cephalosporins [by generation]
119. Cephalexin (Keflex) [1st]
**120. Cefuroxime (Ceftin) [2nd]**
**121. Cefdinir (Omnicef) [3rd]**
122. Ceftriaxone (Rocephin) [3rd]
123. Cefepime (Maxipime) [4th]

Glycopeptide
124. Vancomycin (Vancocin)

## III. Antibiotics – Protein Synthesis Inhibitors (Bacteriostatic)

Tetracyclines
125. Doxycycline (Doryx)
126. Minocycline (Minocin)
**127. Tetracycline (Sumycin)**

Macrolides
128. Azithromycin (Zithromax)
129. Clarithromycin (Biaxin)
130. Erythromycin (E-Mycin)
**131. Fidaxomicin (Dificid)**

Lincosamide
132. Clindamycin (Cleocin)

Oxazolidinone
133. Linezolid (Zyvox)

## IV. Antibiotics - Protein Synth. Inhibitors (Bactericidal)

Aminoglycosides
134. Amikacin (Amikin)
135. Gentamicin (Garamycin)

## V. Antibiotics for Urinary Tract Infections (UTIs) and Peptic Ulcer Disease (PUD)

OTC Urinary tract analgesic
**136. Phenazopyridine (Uristat)**

Nitrofuran
**137. Nitrofurantoin (Macrobid, Macrodantin)**

Dihydrofolate reductase inhibitor
138. Sulfamethoxazole / Trimethoprim (Bactrim DS)

Fluoroquinolones
139. Ciprofloxacin (Cipro)
**140. Gatifloxacin ophthalmic (Zymar)**
141. Levofloxacin (Levaquin)
**142. Moxifloxacin (Avelox) /** [Ophth. is **Vigamox**]

Nitroimidazole
143. Metronidazole (Flagyl)

## VI. Anti-tuberculosis agents

144. Rifampin (Rifadin)
145. Isoniazid (INH)
146. Pyrazinamide (PZA)
147. Ethambutol (Myambutol)

## VII. Antifungals

148. Amphotericin B (Fungizone)
149. Fluconazole (Diflucan)
**150. Ketoconazole (Nizoral)**
151. Nystatin (Mycostatin)

## VIII. Antivirals - Non-HIV

Influenza A and B
152. Oseltamivir (Tamiflu)
153. Zanamivir (Relenza)

Herpes simplex virus & Varicella-Zoster Virus HSV/VSV
154. Acyclovir (Zovirax)
155. Valacyclovir (Valtrex)

Respiratory Syncytial Virus RSV
**156. Palivizumab (Synagis)**

Hepatitis
**157. Entecavir (Baraclude)**
**158. Hepatitis A (Havrix)**
**159. Hepatitis B (Recombivax HB)**

HPV
**160. Human papillomavirus (Gardasil)**

## IX. Antivirals - HIV

Fusion Inhibitor
161. Enfuvirtide (Fuzeon) (T-20)

CCR5 Antagonist
162. Maraviroc (Selzentry) (MVC)

Non-nucleoside reverse transcriptase inhibitors (NNRTI) with two nucleoside / nucleotide reverse transcriptase inhibitors (NRTIs)
**163. Efavirenz (Sustiva) [NNRTI]**
**164. Emtricitabine / Tenofovir (Truvada) [NRTIs]**
165. Efavirenz / Emtricitabine / Tenofovir (Atripla) [NNRTI / NRTIs] (EFV / FTC / TDF)

Integrase Strand Transfer Inhibitor
166. Raltegravir (Isentress) (RAL)

Protease Inhibitor
**167. Atazanavir (Reyataz) (ATV)**
168. Darunavir (Prezista) (DRV)

**X. Miscellaneous**

**169. Albendazole (Albenza) [Anthelmintic]**
**170. Hydroxychloroquine (Plaquenil) [Antimalarial]**
**171. Nitazoxanide (Alinia) [Antiprotozoal]**

## Discussion

I added the prescription antibiotic, **mupirocin** (**Bactroban**) for impetigo. Then I alphabetically listed the antifungal **terbinafine (Lamisil)** after **butenafine (Lotrimin Ultra)**. By adding **betamethasone**, a steroid, to **clotrimazole,** an OTC antifungal, the combination becomes prescription-only **Lotrisone**. I listed the vaccinations around the influenza vaccine. Some require a prescription (depending on varying state laws) and some are antibacterial. I alphabetized them by generic name: **diphtheria toxoid (Boostrix**); ***Haemophilus influenzae* Type B (Pedvax HIB); influenza vaccine (Fluzone, Flumist**); **Measles, Mumps** and **Rubella**

**(MMR)**; **Meningococcal**, both **conjugate and polysaccharide**, **(Menomune)**; **Pertussis in combination**; **Pneumococcal**, both **conjugate and polysaccharide** **(Prevnar 13**, **Pneumovax 23)**, **Polio**, **Rotavirus (RotaTeq)**; and **Tetanus in combination**. Note, **Varivax** and **Zostavax** prevent varicella (chickenpox) and herpes zoster (shingles) respectively. The "vax" in the brand names indicates "vaccine."

**Penicillin** (**Veetids**) follows **amoxicillin** alphabetically. **Cefuroxime** (**Ceftin**), a second-generation cephalosporin, fits in with **cefdinir** (**Omnicef**), a third-generation cephalosporin

**Tetracycline (Sumycin),** although it's essentially unavailable, would follow alphabetically after **doxycycline** and **minocycline**. The "dax" in the name **fidaxomicin (Dificid)**
might come from its source, *Dactylosporangium aurantiacum*. The brand name **Dificid** indicates its primary therapeutic use against *Clostridium difficile*.

**Phenazopyridine (Uristat)** is an over-the-counter urinary tract analgesic that allows a patient to get some relief before she can see her physician for treatment of a bladder infection. **Nitrofurantoin (Macrobid, Macrodantin)** is a nitrofuran antibiotic, as the first brand name implies, and is taken twice daily as indicated by the "b-i-d" in the name. I alphabetically added **gatifloxacin ophthalmic (Zymar)** and **moxifloxacin (Avelox, Vigamox). Vigamox** is an ophthalmic preparation, and contains "v-i" for vision. The antifungal **ketoconazole (Nizoral)** fits alphabetically with the other "azole" antifungal **fluconazole** (**Diflucan**).

**Entecavir (Baraclude)** is for active hepatitis infections, while both **Hepatitis A (Havrix)** and **Hepatitis B**

**(Recombivax HB)** vaccines are preventative. **Gardasil** guards against human papillomavirus (HPV).
I put the HIV medications in category order. **Efavirenz (Sustiva)** is an NNRTI, and both **emtricitabine** and **tenofovir (Truvada)** are nucleotide reverse transcriptase inhibitors (NRTIs). Those three medications are the same medications as are in **Atripla**. The protease inhibitor **atazanavir (Reyataz)** fits in alphabetically.

I finish the immune section with a miscellaneous group I alphabetized by generic name. **Albendazole (Albenza)** is an anthelmintic, which means "against worms;" **hydroxychloroquine (Plaquenil)** is an antimalarial; and **nitazoxanide (Alinia)** is an antiprotozoal.

# Chapter 5 – Neuro

## Medications

### I. OTC Local anesthetics and antivertigo

Local anesthetics
172. Benzocaine (Anbesol) [Ester type]
173. Lidocaine (Solarcaine) [Amide type]

Antivertigo
174. Meclizine (Dramamine, Antivert [RX])

### II. Sedative-hypnotics (Sleeping pills)

OTC Non-narcotic analgesic / Sedative-hypnotic
175. Acetaminophen PM (Tylenol PM)

Benzodiazepine-like
176. Eszopiclone (Lunesta)
177. Zolpidem (Ambien)

Melatonin receptor agonist
178. Ramelteon (Rozerem)

Miscellaneous
**179. Trazodone (Desyrel)**

### III. Antidepressants

Miscellaneous / SSRI
**180. Vilazodone (Viibryd)**

Selective serotonin reuptake inhibitors (SSRIs)
181. Citalopram (Celexa)

182. Escitalopram (Lexapro)
183. Sertraline (Zoloft)
184. Fluoxetine (Prozac, Sarafem)
185. Paroxetine (Paxil, Paxil CR)

Serotonin-Norepinephrine reuptake inhibitors (SNRIs)
186. Duloxetine (Cymbalta)
**187. Desvenlafaxine (Pristiq)**
188. Venlafaxine (Effexor)

Tricyclic antidepressants (TCAs)
189. Amitriptyline (Elavil)
**190. Doxepin (Sinequan)**
**191. Nortriptyline (Pamelor)**

Tetracycylic antidepressant (TeCA) Noradrenergic and specific serotonergic antidepressants (NaSSAs)
**192. Mirtazapine (Remeron)**

Monoamine oxidase inhibitor (MAOI)
193. Isocarboxazid (Marplan)

## IV. Smoking Cessation

194. Bupropion (Wellbutrin, Zyban)
195. Varenicline (Chantix)

## V. Barbiturates

196. Phenobarbital (Luminal)

## VI. Benzodiazepines

197. Alprazolam (Xanax)
198. Midazolam (Versed)

199. Clonazepam (Klonopin)
200. Lorazepam (Ativan)
**201. Temazepam (Restoril)**

## VII. Non-benzodiazepine / non-barbiturate

**202. Buspirone (Buspar)**

## VIII. ADHD medications

Stimulant - Schedule II
**203. Amphetamine/Dextroamphetamine (Adderall)**
204. Dexmethylphenidate (Focalin)
**205. Lisdexamfetamine (Vyvanse)**
206. Methylphenidate (Concerta)

Non-stimulant - non-scheduled
207. Atomoxetine (Strattera)

## IX. Bipolar Disorder

**Simple salt**
208. Lithium (Lithobid)

## XI. Schizophrenia

First generation antipsychotic (FGA) (low potency)
209. Chlorpromazine (Thorazine)

First generation antipsychotic (FGA) (high potency)
210. Haloperidol (Haldol)

Second-generation antipsychotic (SGA)
**211. Aripiprazole (Abilify)**
**212. Olanzapine (Zyprexa)**
213. Risperidone (Risperdal)

214. Quetiapine (Seroquel)

## XII. Antiepileptics

Traditional antiepileptics
215. Carbamazepine (Tegretol)
216. Divalproex (Depakote)
217. Phenytoin (Dilantin)

Newer antiepileptics
218. Gabapentin (Neurontin)
**219. Lamotrigine (Lamictal)**
**220. Levetiracetam (Keppra)**
**221. Oxcarbazepine (Trileptal)**
222. Pregabalin (Lyrica)
**223. Topiramate (Topamax)**
**224. Ziprasidone (Geodon)**

## XIII. Parkinson's, Alzheimer's, Motion sickness

Parkinson's
**225. Benztropine mesylate (Cogentin)**
226. Levodopa / Carbidopa (Sinemet)
227. Selegiline (Eldepryl)
**228. Pramipexole (Mirapex ER)**
**229. Ropinirole (Requip, Requip XL)**

Alzheimer's
230. Donepezil (Aricept)
231. Memantine (Namenda)

Motion sickness
232. Scopolamine (Transderm-Scop)

## DISCUSSION

In practice, **trazodone (Desyrel)** helps patients sleep. **Vilazodone (Viibryd)** is newer, launched in 2011, and is a selective serotonin reuptake inhibitor with partial agonism at the 5-$HT_{1A}$ receptor. **Desvenlafaxine (Pristiq)** is the enantiomer of **venlafaxine (Tranxene)**. I added the tricyclic antidepressants (TCAs) **doxepin (Sinequan)** and **nortriptyline (Pamelor)** before the tetracyclic antidepressant (TeCA) **mirtazapine (Remeron). Remeron** is a noradrenergic and specific serotonergic antidepressant (NaSSA).

I put the barbiturate **phenobarbital (Luminal)** before benzodiazepines because, chronologically, it came earlier and was a more dangerous predecessor. **Temazepam (Restoril)** is a benzodiazepine marketed for sleep disorders, i.e. for "rest" or "restoration." Just as nonfiction literature is classified as "not" fiction, **buspirone (Buspar)** is classified as a non-barbiturate, non–benzodiazepine.

I alphabetized the ADHD medications: **Amphetamine** with **dextroamphetamine (Adderall)** and **Lisdexamfetamine (Vyvanse)**, which has a British "–fetamine" ending. I added two second-generation antipsychotics, **aripiprazole (Abilify)** and **olanzapine (Zyprexa). Abilify** helps a schizophrenic have more "ability to function in society." The WHO "–piprazole" stem is discouraged because it has the PPI "–prazole" stem in it. Newer antiepileptics include **lamotrigine (Lamictal),** which has "ictal," meaning seizure. **Levetiracetam (Keppra), Oxcarbazepine (Trileptal), topiramate (Topamax), ziprasidone (Geodon)** are newer antiseizure meds. The Parkinson's medication **benztropine mesylate (Cogentin)** hints at cognition. **Pramipexole (Mirapex)** twists around the generic name. **Ropinirole (Requip)** "equips" a patient to deal with Parkinson's.

# Chapter 6 – Cardio

## Medications

### I. OTC Antihyperlipidemics and antiplatelet

Antihyperlipidemics
233. Omega-3-acid ethyl esters (Lovaza)
234. Niacin (Niaspan ER)

Antiplatelet
235. Aspirin (Ecotrin)

### II. Diuretics

Osmotic
236. Mannitol (Osmitrol)

Loop
237. Furosemide (Lasix)

Thiazide
238. Hydrochlorothiazide (Microzide)

Potassium sparing and thiazide
239. Triamterene/Hydrochlorothiazide (Dyazide)

Potassium sparing
240. Spironolactone (Aldactone)

Electrolyte replenishment
241. Potassium chloride (K-DUR)

## III. Understanding the Alphas and Betas

Alpha-1 antagonist
242. Doxazosin (Cardura)
**243. Terazosin (Hytrin)**

Alpha-2 agonist
244. Clonidine (Catapres)

Beta-blocker - 1st-generation - non-beta selective
245. Propranolol (Inderal)

Beta-blockers - 2nd-generation - beta selective
246. Atenolol (Tenormin)
**247. Atenolol / Chlorthalidone (Tenoretic)**
**248. Bisoprolol / Hydrochlorothiazide (Ziac)**
249. Metoprolol succinate (Toprol-XL)
250. Metoprolol tartrate (Lopressor)

Beta-blocker - 3rd-generation - non-beta selective vasodilating
251. Carvedilol (Coreg)
**252. Labetalol (Normodyne)**
**253. Nebivolol (Bystolic)**

## IV. Renin-angiotensin-aldosterone system (RAAS)

ACE Inhibitors (ACEIs)
**254. Benazepril / HCTZ (Lotensin HCT)**
255. Enalapril (Vasotec)
**256. Fosinopril (Monopril)**
**257. Quinapril (Accupril)**
258. Lisinopril (Zestril)
**259. Lisinopril / Hydrochlorothiazide (Zestoretic)**
**260. Ramipril (Altace)**

Angiotensin II receptor blockers (ARBs)
**261. Candesartan (Atacand)**
**262. Irbesartan (Avapro)**
**263. Irbesartan / Hydrochlorothiazide (Avalide)**
264. Losartan (Cozaar)
**265. Losartan / Hydrochlorothiazide (Hyzaar)**
266. Olmesartan (Benicar)
**267. Olmesartan / HCTZ (Benicar HCT)**
**268. Telmisartan / HCTZ (Micardis HCT)**
269. Valsartan (Diovan)
**270. Valsartan / HCTZ (Diovan HCT)**

**V. Calcium channel blockers (CCBs)**

Non-dihydropyridines
271. Diltiazem (Cardizem)
272. Verapamil (Calan)

Dihydropyridines
273. Amlodipine (Norvasc)
**274. Amlodipine / Atorvastatin (Caduet)**
**275. Amlodipine / Benazepril (Lotrel)**
**276. Amlodipine / Valsartan (Exforge)**
**277. Felodipine (Plendil)**
278. Nifedipine (Procardia)

**VI. Vasodilators**

**279. Hydralazine (Apresoline)**
**280. Isosorbide mononitrate (Imdur)**
281. Nitroglycerin (Nitrostat)

**VII. Anti-anginal**

**282. Ranolazine (Ranexa)**

## VIII. Antihyperlipidemics

HMG-CoA reductase inhibitors
283. Atorvastatin (Lipitor)
**284. Lovastatin (Mevacor)**
**285. Pravastatin (Pravachol)**
286. Rosuvastatin (Crestor)
**287. Simvastatin (Zocor)**

Fibric acid derivatives
288. Fenofibrate (Tricor)
**289. Gemfibrozil (Lopid)**

Bile acid sequestrant
**290. Colesevelam (Welchol)**

Cholesterol absorption blocker
**291. Ezetimibe (Zetia)**
**292. Ezetimibe / Simvastatin (Vytorin)**

## IX. Anticoagulants and antiplatelets

Anticoagulants
293. Enoxaparin (Lovenox)
294. Heparin
295. Warfarin (Coumadin)
296. Dabigatran (Pradaxa)
**297. Rivaroxaban (Xarelto)**
**298. Apixaban (Eliquis)**

Antiplatelet
**299. Aspirin / Dipyridamole (Aggrenox)**
300. Clopidogrel (Plavix)
**301. Prasugrel (Effient)**
**302. Ticagrelor (Brilinta)**

**X. Cardiac glycoside and Anticholinergic**

Cardiac glycoside
303. Digoxin (Lanoxin)

Anticholinergic
304. Atropine (AtroPen)

**XI. Antidysrhythmic**

**305. Amiodarone (Cordarone)**

## Discussion

The cardio drugs are mostly combinations of old drugs. The alpha-1 antagonist **terazosin (Hytrin)** has an "–azosin" ending and **Hytrin** takes six letters from "hypertension." **Atenolol** with **chlorthalidone** is **Tenoretic**. **Tenoretic** takes the "T-e-n-o-r" from **Tenormin** and adds "r-e-t-i-c" from "diuretic." **Bisoprolol (Zebeta)** and **hydrochlorothiazide (Microzide)** combine to form **Ziac**. **Labetalol (Normodyne)** has "beta" in the generic name. Instead of "–olol" for beta-blocker, the stem is –alol for alpha / beta-blocker. **Nebivolol's** brand name **(Bystolic)** takes letters from systolic (the top blood pressure number) and diastolic (the bottom number).

I grouped renin angiotensin aldosterone system (RAAS) drugs in alphabetical order. For drugs that have a second component, I put the single drug first. I won't go into the medications because by learning "–pril," "-sartan," and "-thiazide," you will know what each drug from 254 to 270 is for.

**Benazepril (Lotensin)** becomes **Lotensin HCT** when the manufacturer adds **hydrochlorothiazide. Fosinopril (Monopril)** and **Quinapril (Accupril)** are unusual in that the ACE inhibitor stem in both their brand and generic names. When a manufacturer adds **hydrochlorothiazide (Microzide)** to **Lisinopril (Zestril),** it becomes **Zestoretic** by adding the last letters of "diuretic" to the name. The manufacturers of ARBs similarly create brand names for new combination products by adding either "-lide," "Hy-," or "HCT." The calcium channel blocker additions also add nothing that isn't in the original 200: "-dipine" for the CCB dihydropyridine, "-vastatin" for the HMG-CoAs, and "-pril" for the ACEIs.

**Ranolazine (Ranexa)** and **hydralazine (Apresoline)** both have similar endings, but **ranolazine** *treats* angina pectoris and **hydralazine** may *cause* angina pectoris. I grouped **hydralazine** with another vasodilator, **isosorbide dinitrate (Imdur).** The "nitrate" helps you remember this is in the same class as **nitroglycerin**.

I listed the HMG-CoA reductase inhibitors **lovastatin (Mevacor), pravastatin (Pravachol),** and **simvastatin (Zocor)** alphabetically. The brand names hint at coronary or cholesterol. You can see the "fib" in their fibric acid derivative **gemfibrozil (Lopid).** That brand name **Lopid** hints at "lowering lipids." The bile acid sequestrant **colesevelam (Welchol)** has both a generic and brand name with a hint at cholesterol. **Ezetimibe (Zetia)** is a cholesterol absorption blocker, a different type of drug, and can be combined with an HMG-CoA like **simvastatin (Zocor)** to make **Vytorin.**

The anticoagulants **dabigatran (Pradaxa), rivaroxaban (Xarelto)**, and **Apixaban (Eliquis)** are usually coupled because they don't require monitoring like their counterpart, **warfarin (Coumadin). Dipyridamole / aspirin (Aggrenox)** work together to prevent clots and the brand name can be thought of as "aggregate not." **Prasugrel (Effient)** and **Ticagrelor (Brilinta)** share a "-grel" stem with **clopidogrel (Plavix). Effient** is the word "efficient" without the "c-i," so may be efficient at thinning platelets. **Ticagrelor (Brilinta)** has demonstrated superiority to **clopidogrel**. The brand name **Amiodarone (Cordarone)** and the brand and generic share the "arone" lettering. Cardiologists can also use beta-blockers, calcium channel blockers, and **digoxin** as antidysrhythmics.

# CHAPTER 7 – ENDOCRINE / MISC.

## MEDICATIONS

### I. OTC Insulin and emergency contraception

306. Regular Insulin (Humulin R)
307. NPH Insulin (Humulin N)
308. Levonorgestrel (Plan B One-Step)

### II. Diabetes and insulin

Biguanides
309. Metformin (Glucophage)
**310. Metformin / Glyburide (Glucovance)**

DPP-4 Inhibitors (Gliptins)
**311. Linagliptin (Tradjenta)**
**312. Saxagliptin (Onglyza)**
313. Sitagliptin (Januvia)

Meglitinides (Glinides)
**314. Repaglinide (Prandin)**

Sulfonylureas - 2nd-generation
315. Glyburide (DiaBeta)
**316. Glimepiride (Amaryl)**
317. Glipizide (Glucotrol)

Thiazolidinediones (Glitazones)
**318. Pioglitazone (Actos)**
**319. Rosiglitazone (Avandia)**

Incretin mimetics
**320. Exenatide (Byetta)**
**321. Liraglutide (Victoza)**

Hypoglycemia
322. Glucagon (GlucaGen)

RX Insulin
**323. Insulin aspart (Novolog)**
324. Insulin lispro (Humalog)
**325. Insulin detemir (Levemir)**
326. Insulin glargine (Lantus, Toujeo)

**III. Thyroid hormones**

Hypothyroidism
327. Levothyroxine (Synthroid)

Hyperthyroidism
328. Propylthiouracil (PTU)

**IV. Hormones and contraception**

Low testosterone
329. Testosterone (AndroGel)

Estrogens and / or Progestins
330. Estradiol (Estrace, Estraderm)
331. Conjugated estrogens (Premarin)
332. Conjugated estrogens / Medroxyprogesterone (Prempro, Premphase)
333. Progesterone (Prometrium)
334. Medroxyprogesterone (Provera)

Combined oral contraceptive pill (COCP)
335. Ethinyl estradiol / norethindrone / Fe (Loestrin 24 Fe)
336. Ethinyl estradiol / norgestimate (Tri-Sprintec)

Patch
337. Ethinyl estradiol / norelgestromin (OrthoEvra)

Ring
338. Ethinyl estradiol / etonogestrel (NuvaRing)

**V. Overactive bladder, urinary retention, erectile dysfunction (ED), benign prostatic hyperplasia (BPH)**

Overactive bladder
339. Oxybutynin (Ditropan)
**340. Darifenacin (Enablex)**
341. Solifenacin (VESIcare)
342. Tolterodine (Detrol)

Urinary retention
343. Bethanechol (Urecholine)

Erectile dysfunction - PDE-5 inhibitors
344. Sildenafil (Viagra)
**345. Vardenafil (Levitra)**
346. Tadalafil (Cialis)

BPH – Alpha-blocker
347. Alfuzosin (Uroxatral)
348. Tamsulosin (Flomax)

BPH – 5-alpha-reducase inhibitor
349. Dutasteride (Avodart)
350. Finasteride (Proscar, Propecia)

## DISCUSSION

I alphabetized the antidiabetic oral medications by class. Manufacturers combine the biguanide **metformin (Glucophage)** with the sulfonylurea **glyburide (DiaBeta)** to make **Glucovance** – an advance in glucose lowering. Patients refer to DPP-4 inhibitors **linagliptin (Tradjenta)** and **saxagliptin (Onglyza)** by their stem "-gliptin," the meglitinide **repaglinide (Prandin),** as a "-glinide," and the thiazolidinediones **pioglitazone (Actos)** and **rosiglitazone (Avandia),** as "-glitazones." Both incretin mimetics **exenatide (Byetta)** and **liraglutide (Victoza)** are injectables. **Insulin aspart (Novolog)** is another insulin analog. **Insulin detemir (Levemir)** is a long-acting insulin like **Lantus**.

Estrogens first, then combo estrogen / progestin products, then progesterone products, recognizable by their "estr-" and "-gest-" stems respectively. **Estradiol (Estrace, Estraderm)** has brand names with the estrogen stem. **Premarin** is for "pregnant mare's urine," the drug's source. **Prempro** and **Premphase** have different estrogen / progestin levels. **Progesterone (Prometrium)** and **medroxyprogesterone (Provera)** are progestin tablets.

**Darifenacin (Enablex)** and **solifenacin (VESIcare)** have the same "–fenacin" stem. Both work for overactive bladder (OAB). The **Enablex** brand name hints at "enabling" the patient to "exit" the house when the OAB might have kept them in. **Vardenafil (Levitra)** seems to have the same "-den-" infix as **sildenafil (Viagra)**. To levitate is to rise above the ground, so the brand **Levitra** hints at the rising erection.

# APPENDIX

## ANSWERS TO DRUG QUIZZES (LEVEL 1)

**Gastrointestinal drugs**
1. A 2. E 3. D 4. E 5. F 6. B 7. A/D 8. G 9. F 10. C
2. –tidine 4. –tidine 5. –prazole 8. –liximab 9. –prazole
10. -setron

**Musculoskeletal drugs**
1. E 2.C 3. F 4. B 5. D 6. I 7. I 8. G 9. H 10. A
2. –dronate 4. –xostat 5. –nercept 8. –profen 9. –coxib
10. -triptan

**Respiratory drugs**
1. I 2. B 3. A 4. J 5. F 6. C 7. B 8. E 9. D 10. G
1. –terol 4. –terol 6. –tropium 7. –atadine 8. –lukast
9. –drine 10. -pred-

**Immune system drugs**
1. H 2. G 3. D 4. C 5. K 6. E 7. L 8. F 9. K 10. M
1. –cillin 2. –thromycin 3. Cef- 4. Cef- 5. –conazole
6. –micin 8. -floxacin 10. –cyclovir

**Nervous system drugs**
1. K 2. F 3. A 4. E 5. B 6. H 7. J 8. C 9. L 10. M
1. –azolam 2. -triptyline 3. –oxetine 7. –peridol 9. –dopa
10. –pidem

**Cardio system drugs**
1. O 2. E 3. A 4. D 5. K 6. N 7. C 8. F 9. H 10. M
1. –vastatin 2. –grel 3. –pril 4. –parin 5. –semide
6. -thiazide 7. –sartan 8. –olol 9. –dipine

**Endocrine / Misc. system drugs**
1. A 2. H 3. A 4. M 5. I 6. A 7. J 8. N 9. L 10. K
1. Gli- 3. Gly- 6. –formin 9. –fenacin 10. –afil

## ANSWERS TO DRUG QUIZZES (LEVEL 2)

**Gastrointestinal drugs**
1. B 2. F 3. F 4. D 5. C 6. A 7. E 8. D 9. B 10. E
1. -sal- 2. –prazole 3. –prazole 7. –tidine 10. –tidine

**Musculoskeletal drugs**
1. D 2. I 3. D 4. G 5. B 6. C 7. B 8. G 9. C 10. I
1. –trexate 3. –tacept 5. –xostat 6. -dronate 9. -dronate

**Respiratory drugs**
1. J 2. F 3. H 4. J 5. D 6. B 7. A 8. I 9. E 10. G
1. –terol 4. –terol 5. –drine 8. –terol 9. –lukast 10. Pred-

**Immune system drugs**
1. L 2. K 3. E 4. F 5. L 6. M 7. A 8. I 9. G 10. O
1. Rif- 3. –kacin 4. –floxacin 6. –cyclovir 7. Ceph-
8. Sulfa- / -prim 9. –thromycin 10. -amivir

**Nervous system drugs**
1. D 2. K 3. H 4. I 5. C 6. M 7. E 8. E 9. L 10. N
1. –faxine 2. –azepam 3. –toin 4. –tiapine 6. –clone
7. –oxetine 9. –giline

**Cardio system drugs**
1. I 2. G 3. C 4. N 5. B 6. H 7. P 8. A 9. J 10. D
1. –tiazem 2. –dil- (-olol) 3. –sartan 4. –thiazide 5. –azosin
6. –dipine 7. Nitro- 8. –pril 10. –farin

**Endocrine / Misc. system drugs**
1. B 2. C 3. L 4. K 5. E 6. F 7. L 8. O 9. K 10. B
1. –steride 4. –afil 5. estr- 6. -gest- / estr 9. –afil 10. -steride

# ANSWERS TO FINAL EXAM (LEVEL 1)

| | | | | | | | |
|---|---|---|---|---|---|---|---|
| 1 | f | 26 | n | 51 | f | 76 | k |
| 2 | o | 27 | h | 52 | i | 77 | u |
| 3 | g | 28 | c | 53 | n | 78 | i |
| 4 | q | 29 | j | 54 | i | 79 | i |
| 5 | t | 30 | f | 55 | h | 80 | m |
| 6 | r | 31 | i | 56 | q | 81 | g |
| 7 | r | 32 | a | 57 | r | 82 | l |
| 8 | g | 33 | l | 58 | m | 83 | t |
| 9 | m | 34 | l | 59 | f | 84 | l |
| 10 | t | 35 | l | 60 | f | 85 | r |
| 11 | q | 36 | e | 61 | l | 86 | e |
| 12 | q | 37 | d | 62 | m | 87 | k |
| 13 | k | 38 | k | 63 | g | 88 | f |
| 14 | l | 39 | h | 64 | a | 89 | l |
| 15 | p | 40 | g | 65 | b | 90 | a |
| 16 | p | 41 | k | 66 | f | 91 | s |
| 17 | i | 42 | o | 67 | k | 92 | c |
| 18 | s | 43 | j | 68 | l | 93 | p |
| 19 | n | 44 | g | 69 | o | 94 | j |
| 20 | f | 45 | e | 70 | c | 95 | k |
| 21 | d | 46 | l | 71 | f | 96 | g |
| 22 | m | 47 | k | 72 | l | 97 | d |
| 23 | b | 48 | g | 73 | k | 98 | c |
| 24 | a | 49 | f | 74 | p | 99 | d |
| 25 | j | 50 | m | 75 | d | 100 | k |

# ANSWERS TO FINAL EXAM (LEVEL 2)

| | | | | | | | |
|---|---|---|---|---|---|---|---|
| 1 | b | 26 | j | 51 | o | 76 | o |
| 2 | a | 27 | h | 52 | f | 77 | l |
| 3 | k | 28 | i | 53 | j | 78 | r |
| 4 | i | 29 | k | 54 | f | 79 | p |
| 5 | r | 30 | g | 55 | r | 80 | f |
| 6 | p | 31 | c | 56 | d | 81 | b |
| 7 | s | 32 | l | 57 | h | 82 | p |
| 8 | t | 33 | l | 58 | o | 83 | a |
| 9 | f | 34 | a | 59 | c | 84 | d |
| 10 | f | 35 | l | 60 | i | 85 | r |
| 11 | p | 36 | k | 61 | p | 86 | p |
| 12 | d | 37 | j | 62 | f | 87 | b |
| 13 | j | 38 | m | 63 | e | 88 | h |
| 14 | b | 39 | o | 64 | p | 89 | i |
| 15 | q | 40 | e | 65 | h | 90 | n |
| 16 | s | 41 | e | 66 | f | 91 | b |
| 17 | c | 42 | g | 67 | l | 92 | f |
| 18 | e | 43 | d | 68 | k | 93 | q |
| 19 | h | 44 | o | 69 | m | 94 | q |
| 20 | l | 45 | k | 70 | f | 95 | i |
| 21 | q | 46 | f | 71 | j | 96 | d |
| 22 | r | 47 | h | 72 | p | 97 | a |
| 23 | h | 48 | f | 73 | r | 98 | l |
| 24 | q | 49 | l | 74 | b | 99 | f |
| 25 | s | 50 | n | 75 | q | 100 | c |

## ALPHABETICAL LIST OF STEMS

**ac** anti-inflammatory agents (acetic acid derivatives)

**adol** analgesics (mixed opiate receptor agonists/antagonists)

**afil** phosphodiesterase type 5 (PDE5) inhibitors

**alol** combined alpha and beta blockers

**amivir** neuraminidase inhibitors

**astine** antihistaminics (histamine-$H_1$ receptor antagonists)

**atadine** tricyclic histaminic-$H_1$ receptor antagonists, loratadine derivatives (formerly -tadine)

**azepam** antianxiety agents (diazepam type)

**azolam** (WHO stem) diazepam derivatives

**azosin** antihypertensives (prazosin type)

**barb** barbituric acid derivatives

**bendazole** anthelmintics (tibendazole type)

**caine** local anesthetics

**cavir** carbocyclic nucleosides

**cef** cephalosporins

**citabine** nucleoside antiviral / antineoplastic agents, cytarabine or azarabine derivatives

**cillin** penicillins

**clone** hypnotics/tranquilizers (zopiclone type)

**conazole** systemic antifungals (miconazole type)

**coxib** cyclooxygenase-2 inhibitors

**cycline** antibiotics (tetracycline derivatives)

**cyclovir** antivirals (acyclovir type)

**dil** vasodilators (undefined group)

**dipine** phenylpyridine vasodilators (nifedipine type)

**dopa** dopamine receptor agonists

**dralazine** antihypertensives (hydrazine-phthalazines)

**drine** sympathomimetics

| | |
|---|---|
| **dronate** | calcium metabolism regulators |
| **estr** | estrogens |
| **farin** | warfarin analogs |
| **faxine** | antianxiety, antidepressant inhibitor of norepinephrine and dopamine re-uptake |
| **fenacin** | muscarinic receptor antagonists |
| **fetamine** | amfetamine derivatives |
| **fibrate** | antihyperlipidemics (clofibrate type) |
| **floxacin** | fluoroquinolone (not on Stem List) |
| **formin** | hypoglycemics (phenformin type) |
| **gab** | gabamimetics |
| **gatran** | thrombin inhibitors (argatroban type) |
| **gest** | progestins |
| **giline** | Monoamine oxidase (MAO) inhibitors, type B |
| **gli (was gly)** | antihyperglycemics |
| **glinide** | antidiabetic, sodium glucose co-transporter 2 (SGLT2) inhibitors, not phlorozin derivatives |
| **gliptin** | dipeptidyl aminopeptidase-IV inhibitors |
| **glitazone** | peroxisome proliferator activating receptor (PPAR) agonists (thiazolidene derivatives) |
| **glutide** | glucagon-like peptide (GLP) analogs |
| **gly** | antihyperglycemics |
| **grel** | platelet aggregation inhibitors, primarily platelet P2Y12 receptor antagonists |
| **icam** | anti-inflammatory agents (isoxicam type) |
| **ifene** | antiestrogens of the clomifene and tamoxifen groups |
| **imibe** | antihyperlipidaemics, acyl CoA: cholesterol acyltransferase (ACAT) inhibitors |
| **iodarone** | indicates high iodine content antiarrhythmic |
| **kacin** | antibiotics obtained from *Streptomyces kanamyceticus* (related to kanamycin) |

**liximab** monoclonal antibodies

**lizumab** monoclonal antibodies

**lukast** leukotriene receptor antagonists

**mantine** antivirals/antiparkinsonians (adamantane derivatives)

**melteon** selective melatonin receptor agonist

**methacin** anti-inflammatory agents (indomethacin type)

**micin** antibiotics (*Micromonospora* strains)

**mycin** antibiotics (*Streptomyces* strain)

**nal** narcotic agonists/antagonists (normorphine type)

**navir** HIV protease inhibitors (saquinavir type)

**nercept** tumor necrosis factor receptors

**nicline** nicotinic acetylcholine receptor partial agonists/agonists

**nidazole** antiprotozoal substances (metronidazole type)

**nifur** 5-nitrofuran derivatives

**nitro** (WHO stem) $NO_2$ derivatives

**olol** beta-blockers (propranolol type)

**orphan** narcotic antagonists/agonists (morphinan derivatives)

**oxacin** antibacterials (quinolone derivatives)

**oxanide** antiparasitics (salicylanilide derivatives)

**oxetine** antidepressants (fluoxetine type)

**pamil** coronary vasodilators (verapamil type)

**parin** heparin derivatives and low molecular weight (or depolymerized) heparins

**peg** PEGylated compounds, covalent attachment of macrogol (polyethylene glycol) polymer

**peridol** antipsychotics (haloperidol type)

**peridone** antipsychotics (risperidone type)

**pezil** acetylcholinesterase inhibitors used in the treatment of Alzheimer's disease

**pidem** hypnotics/sedatives (zolpidem type)

**pin(e)** tricyclic compounds

**piprazole** (WHO stem) psychotropics, phenylpiperazine derivatives (future use is discouraged due to conflict with stem -prazole)

**prazole** antiulcer agents (benzimidazole derivatives)

**pred** prednisone and prednisolone derivatives

**pril** antihypertensives (ACE inhibitors)

**prim** antibacterials (trimethoprim type)

**profen** anti-inflammatory/analgesic agents (ibuprofen type)

**prost** prostaglandins

**racetam** nootropic agents (learning, cognitive enhancers), piracetam type

**rifa** antibiotics (rifamycin derivatives)

**sal** anti-inflammatory agents (salicylic acid derivatives)

**sartan** angiotensin II receptor antagonists

**semide** diuretics (furosemide type)

**setron** serotonin $5\text{-HT}_3$ receptor antagonists

**sidone** antipsychotic with binding activity on serotonin (5-HT2A) and dopamine (D2) receptors

**spirone** anxiolytics (buspirone type)

**ster** steroids (androgens, anabolics)

**steride** testosterone reductase inhibitors

**sulfa** antimicrobials (sulfonamides derivatives)

**tacept** T-cell receptors

**tegravir** integrase inhibitors

**terol** bronchodilators (phenethylamine derivatives)

**thiazide** diuretics (thiazide derivatives)

**thromycin** macrolide (not on Stem List)

**tiapine** antipsychotics (dibenzothiazepine derivatives)

**tiazem** calcium channel blockers (diltiazem type)

| | |
|---|---|
| **tide** | peptides |
| **tidine** | $H_2$-receptor antagonists (cimetidine type) |
| **toin** | antiepileptics (hydantoin derivatives) |
| **traline** | selective serotonin reuptake inhibitors (SSRI) |
| **trexate** | antimetabolites (folic acid derivatives) |
| **triptan** | antimigraine agents (5-$HT_1$ receptor agonists); sumatriptan derivatives |
| **triptyline** | antidepressants (dibenzol[a.d.]cycloheptane derivatives) |
| **trop(ium)** | atropine derivative, (quaternary ammonium salt) |
| **trop(ine)** | atropine derivatives; Subgroups: tertiary amines (e.g., benztropine) |
| **uracil** | uracil derivatives used as thyroid antagonists and as antineoplastics |
| **vastatin** | antihyperlipidemics (HMG-CoA inhibitors) |
| **vir** | antivirals |
| **virenz** | non-nucleoside reverse transcriptase inhibitors; benzoxazinone derivatives |
| **viroc** | CC chemokine receptor type 5 (CCR5) antagonists |
| **vudine** | antineoplastics; antivirals (zidovudine group) (exception: edoxudine) |
| **xaban** | antithrombotics, blood coagulation factor XA inhibitors |
| **xostat** | xanthine oxidase/dehydrogenase inhibitors |
| **zolid** | oxazolidinone antibacterials |

# LIST OF STEMS BY PHYSIOLOGIC CLASS

## Chapter 1: Gastrointestinal

| | |
|---|---|
| **liximab** | monoclonal antibodies |
| **peg** | PEGylated compounds, covalent attachment of macrogol (polyethylene glycol) polymer |
| **prazole** | antiulcer agents (benzimidazole derivatives) |
| **prost** | prostaglandins |
| **sal** | anti-inflammatory agents (salicylic acid derivatives) |
| **setron** | serotonin 5-$HT_3$ receptor antagonists |
| **tidine** | $H_2$-receptor antagonists (cimetidine type) |

## Chapter 2: Musculoskeletal

| | |
|---|---|
| **ac** | anti-inflammatory agents (acetic acid derivatives) |
| **adol** | analgesics (mixed opiate receptor agonists/antagonists) |
| **coxib** | cyclooxygenase-2 inhibitors |
| **dronate** | calcium metabolism regulators |
| **icam** | anti-inflammatory agents (isoxicam type) |
| **ifene** | antiestrogens of the clomifene and tamoxifen groups |
| **liximab** | monoclonal antibodies |
| **methacin** | anti-inflammatory agents (indomethacin type) |
| **nal** | narcotic agonists/antagonists (normorphine type) |
| **nercept** | tumor necrosis factor receptors |
| **profen** | anti-inflammatory/analgesic agents (ibuprofen type) |
| **tacept** | T-cell receptors |
| **trexate** | antimetabolites (folic acid derivatives) |
| **triptan** | antimigraine agents (5-$HT_1$ receptor agonists); sumatriptan derivatives |
| **xostat** | xanthine oxidase/dehydrogenase inhibitors |

## Chapter 3: Respiratory

**atadine** tricyclic histaminic-$H_1$ receptor antagonists, loratadine derivatives (formerly -tadine)
**astine** antihistaminics (histamine-$H_1$ receptor antagonists)
**drine** sympathomimetics
**lizumab** monoclonal antibodies
**lukast** leukotriene receptor antagonists
**orphan** narcotic antagonists/agonists (morphinan derivatives)
**pred** prednisone and prednisolone derivatives
**terol** bronchodilators (phenethylamine derivatives)
**trop(ium)** atropine derivative (quaternary ammonium salt)

## Chapter 4: Immune

**amivir** neuraminidase inhibitors
**bendazole** anthelmintics (tibendazole type)
**cavir** carbocyclic nucleosides
**cef** cephalosporins
**cillin** penicillins
**citabine** nucleoside antiviral / antineoplastic agents, cytarabine or azarabine derivatives
**conazole** systemic antifungals (miconazole type)
**cycline** antibiotics (tetracycline derivatives)
**cyclovir** antivirals (acyclovir type)
**floxacin** fluoroquinolone (not on Stem List)
**kacin** antibiotics obtained from *Streptomyces kanamyceticus* (related to kanamycin)
**lizumab** monoclonal antibodies
**micin** antibiotics (*Micromonospora* strains)
**mycin** antibiotics (*Streptomyces* strain)
**navir** HIV protease inhibitors (saquinavir type)
**nidazole** antiprotozoal substances (metronidazole type)
**nifur** 5-nitrofuran derivatives
**oxacin** antibacterials (quinolone derivatives)
**oxanide** antiparasitics (salicylanilide derivatives)
**prim** antibacterials (trimethoprim type)

**rifa** antibiotics (rifamycin derivatives)
**sulfa** antimicrobials (sulfonamides derivatives)
**tegravir** integrase inhibitors
**thromycin** macrolide (not on Stem List)
**vir** antivirals
**virenz** non-nucleoside reverse transcriptase inhibitors; benzoxazinone derivatives
**viroc** CC chemokine receptor type 5 (CCR5) antagonists
**vudine** antineoplastics; antivirals (zidovudine group) (exception: edoxudine)
**zolid** oxazolidinone antibacterials

## Chapter 5: Neuro

**azepam** antianxiety agents (diazepam type)
**azolam** (WHO stem) diazepam derivatives
**caine** local anesthetics
**clone** hypnotics/tranquilizers (zopiclone type)
**dopa** dopamine receptor agonists
**faxine** antianxiety, antidepressant inhibitor of norepinephrine and dopamine re-uptake
**gab** gabamimetics
**giline** Monoamine oxidase (MAO) inhibitors, type B
**melteon** selective melatonin receptor agonist
**oxetine** antidepressants (fluoxetine type)
**peridol** antipsychotics (haloperidol type)
**peridone** antipsychotics (risperidone type)
**pezil** acetylcholinesterase inhibitors used in the treatment of Alzheimer's disease
**pidem** hypnotics/sedatives (zolpidem type)
**pin(e)** tricyclic compounds
**tiapine** antipsychotics (dibenzothiazepine derivatives)
**toin** antiepileptics (hydantoin derivatives)
**traline** selective serotonin reuptake inhibitors (SSRI)
**triptyline** antidepressants (dibenzol[a,d] cycloheptane derivatives)
**nicline** nicotinic acetylcholine receptor partial agonists/agonists

**barb** barbituric acid derivatives
**spirone** anxiolytics (buspirone type)
**fetamine** amfetamine derivatives
**piprazole** (WHO stem) psychotropics, phenylpiperazine derivatives (future use is discouraged due to conflict with stem -prazole)
**racetam** nootropic agents (learning, cognitive enhancers), piracetam type
**sidone** antipsychotic with binding activity on serotonin (5-HT2A) and dopamine (D2) receptors
**tropine** atropine derivatives; Subgroups: tertiary amines (e.g., benztropine)
**mantine** antivirals/antiparkinsonians (adamantane derivatives)

## Chapter 6: Cardio

**alol** combined alpha and beta blockers
**azosin** antihypertensives (prazosin type)
**dil** vasodilators (undefined group)
**dipine** phenylpyridine vasodilators (nifedipine type)
**dralazine** antihypertensives (hydrazine-phthalazines)
**farin** warfarin analogs
**fibrate** antihyperlipidemics (clofibrate type)
**gatran** thrombin inhibitors (argatroban type)
**grel** platelet aggregation inhibitors, primarily platelet P2Y12 receptor antagonists
**imibe** antihyperlipidaemics, acyl CoA: cholesterol acyltransferase (ACAT) inhibitors
**iodarone** indicates high iodine content antiarrhythmic
**nitro** (WHO stem) $NO_2$ derivatives
**olol** beta-blockers (propranolol type)
**pamil** coronary vasodilators (verapamil type)
**parin** heparin derivatives and low molecular weight (or depolymerized) heparins
**pril** antihypertensives (ACE inhibitors)
**sartan** angiotensin II receptor antagonists
**semide** diuretics (furosemide type)
**thiazide** diuretics (thiazide derivatives)

| | |
|---|---|
| **tiazem** | calcium channel blockers (diltiazem type) |
| **trop(ine)** | atropine derivatives; Subgroups: tertiary amines (e.g., benztropine) |
| **vastatin** | antihyperlipidemics (HMG-CoA inhibitors) |
| **xaban** | antithrombotics, blood coagulation factor XA inhibitors |

## Chapter 7: Endocrine / Misc.

| | |
|---|---|
| **afil** | phosphodiesterase type 5 (PDE5) inhibitors |
| **estr** | estrogens |
| **fenacin** | muscarinic receptor antagonists |
| **formin** | hypoglycemics (phenformin type) |
| **gest** | progestins |
| **gli (was gly)** | antihyperglycemics |
| **glinide** | antidiabetic, sodium glucose co-transporter 2 (SGLT2) inhibitors, not phlorozin derivatives |
| **gliptin** | dipeptidyl aminopeptidase-IV inhibitors |
| **glitazone** | peroxisome proliferator activating receptor (PPAR) agonists (thiazolidene derivatives) |
| **glutide** | glucagon-like peptide (GLP) analogs |
| **gly** | antihyperglycemics |
| **ster** | steroids (androgens, anabolics) |
| **steride** | testosterone reductase inhibitors |
| **tide** | peptides |
| **uracil** | uracil derivatives used as thyroid antagonists and as antineoplastics |

# GENERIC AND BRAND NAME INDEX